Remote Sensing and its Applications

This Publication is supported by the Department of Space

Remote Sensing and its Applications

L R A NARAYAN

Universities Press

Universities Press (India) Private Limited

Registered Office
3-6-747/1/A & 3-6-754/1, Himayatnagar, Hyderabad 500 029 (A.P.), India
e-mail: info@universitiespress.com

Distributed by
Orient Blackswan Private Limited

Registered Office
3-6-752, Himayatnagar, Hyderabad 500 029 (A.P.), India

Other Offices
Bangalore / Bhopal / Bhubaneshwar / Chennai / Ernakulam
Guwahati / Hyderabad / Jaipur / Kolkata / Lucknow / Mumbai
New Delhi / Patna

First Published 1999
Reprinted 2001, 2011, 2012 2013

ISBN: 978 81 7371 268 5

Typeset by
Ventre Graphics Pvt. Ltd., Chennai 600 031

Printed in India at
Charita Impressions, Hyderabad

Published by
Universities Press (India) Private Limited
3-6-747/1/A & 3-6-754/1, Himayatnagar
Hyderabad 500 029 (A.P), India

Contents

Foreword

Space-based remote sensing is a comparatively new technology but with high potential for applications in almost all facets of resources management. India recognised the potential of this technology quite early. It had conducted several remote sensing experiments in the late, '60s and '70s using airborne sensors and receiving data from the US Landsat.

The experimental Earth observation satellites of India, Bhaskara-1 and Bhaskara-II, launched in the beginning of the '80s, gave further fillip to the advancement of remote sensing in the country. Today, India is in the forefront, in terms of building the world's best Indian Remote Sensing satellites like IRS-1C and IRS-1D and also in the application of the data received from these satellites in important fields like agriculture, forestry, water harvesting, mineral prospecting, urban planning, land use and land cover planning, etc. India has also taken up the Integrated Mission for Sustainable Development (IMSD) which makes use of space-based remote sensing data along with socio-economic data.

In spite of many efforts undertaken in the field of remote sensing, there has been a lacuna in making the general public aware of this technology and its great potential. I am extremely happy that Dr L R A Narayan, who has been associated with remote sensing technology from its early days, has taken on the task of bringing about an awareness on remote sensing through his writings in *The Hindu* and now by publishing a book.

Through this book, I am sure that the readers will be able to appreciate many aspects of remote sensing technology and its potential for various applications.

K Kasturirangan

Chairman, Space Commission
and Secretary to the
Govt. of India, Dept. of Space

Preface

The motivation for me to write these articles came from the discussion I had with the Editor and Associate Editor of *The Hindu*, Mr N Ravi and Mrs Nirmala Lakshman, who happened to see a magazine article titled "The Reporters in the Sky" which referred to the various activities that preluded the Gulf War and to some extent during and after the war in 1992, based on observations from satellites with Remote Sensing Sensors. I met Prof U R Rao, the Chairman of Indian Space Research Organisation, (ISRO), with a letter from the Editor of *The Hindu* after having published my first one or two articles, in November 1993. Prof Rao was generous and encouraged me to write as much as I could and asked the ISRO centres to make available all the information that I needed from time to time.

I thought of stopping writing after publishing some half-a-dozen articles, but the tremendous reader responses demanded that I continue. Therefore, it is my primary responsibility to acknowledge my sincere gratitude and thanks for the unstinting support that I received from *The Hindu* and the ISRO Headquarters.

The majority of the articles are based on various projects carried out by the National Remote Sensing Agency, Hyderabad; the Space Application Centre, ISRO, Ahmedabad; the Regional Remote Sensing Service Centres of ISRO at Bangalore, Nagpur, Dehradun, Jodhpur, Kharagpur and organisations like ISRO Satellite Centre at Bangalore; Institute of Remote Sensing, Anna University; Indian Institute of Remote Sensing (IIRS), Dehradun; the J N T University, Hyderbad and also from the NNRMS and ISRO Headquarters at Bangalore. I must thank from the bottom of my heart all the heads of these organisations and the scientists who supported my venture, wholeheartedly. In addition, I had cooperation and help from the Andhra Pradesh State Remote Sensing Application Centre, Hyderabad, the Maharashtra State Remote Sensing Application centre and several other state Remote Sensing centres, all of whom I am unable to acknowledge individually here.

Being a Distinguished Fellow during the last six years in the M S Swaminathan Research Foundation in Chennai, an internationally reputed non-government organisation gave me an insight into how science and technology cannot develop without its commitment to the society. I feel extremely lucky to be associated with Prof M S Swaminathan, a towering international yet down-to-earth scientist, and the continuous encouragement

I received all these years and continue to do so. My wholehearted pranams to this great personality, the equal to whom I am yet to come across, as but for his vision and mission India's food security would have been in great jeopardy.

I have had many requests to compile these articles in book form, avoiding as much technical jargon as possible, and thus penetrating bit by bit into some of the universities and schools in their curriculum at different levels, as many students and teachers wanted an easily understandable and affordable book for their day-to-day reference. The general public, who had very little information on this topic, also showed enthusiasm.

I will be failing in my duty if I do not mention Dr K Kasturirangan, the present Chairman ISRO/NRSA and Secretary to Government of India, Department of Space, who came forward to encourage me by agreeing to purchase an appreciable number of copies. My sincere thanks are due to him. I am also grateful to Shri R Venkatesan, IA & AS (Retd.) who was Additional Secretary, in the Department of Space, and later Secretary to the Govt. of India and Dr S Krishnamurthy, Director of Publications, ISRO, Bangalore, for their timely help and co-operation in securing the ISRO support.

Ultimately, Dr G Thyagarajan, Scientific Secretary, COSTED (Committee on Science and Technology in Developing Countries) and also the President of the Madras Science Foundation with his scientists and staff, in particular Dr B Babuji, came forward and undertook to get this published, through a reputed publisher as they felt that reading material in this subject is necessary not only in India but also in a large number of developing countries, with whom COSTED has links. But for COSTED, I would not have got this book published so efficiently. I would like to express my sincere gratitude to Dr Thyagarajan, Dr. Babuji and the entire staff of COSTED, Chennai for help rendered. The Universities Press (India) Limited, Hyderabad, one of the well known scientific book publishers known nationally and internationally, was very generous in doing all that they could do to bring out the book in a very attractive form and ensure its distribution all over the country and abroad. I am immensely grateful fo them.

Finally, as I mentioned earlier, I have acknowledged as many people whom I could remember; however, in case I have omitted anyone, I tender my apologies in advance.

There are 47 articles in this book, all published in *The Hindu*, at the rate of about one article a month in the Sunday colour supplement, during the past six years. I feel encouraged to continue writing even after the publication of this book as long as I can, and perhaps, if a second edition is warranted at a later date, my subsequent articles will find place in it.

Remote Sensing and Geographical Information System, with support from the Global Positioning System, have been perhaps the greatest technological breakthrough in the last quarter of this century. They are expected to make greater breakthroughs in the 21st century, to manage our natural resources more successfully, so that sustainability is maintained in spite of growing population and shrinking resources.

I might have made some unintentional errors, I hope the readers and critics would be tolerant and pardon me for the lapses. I welcome constructive and critical suggestions. I must also state here that the articles presented here are not ordered according to any particular sequence. One final word is that individual acknowledgements by names could not be made, as it would have been impossible to do, only the organizations to which they belong have been mentioned above. I beg pardon for this omission.

September 1999 — L R A NARAYAN

Introduction

Remote Sensing is a revolutionary new tool that lends itself to useful applications in a variety of areas of high relevance to every country in the world. The civilian application areas include Agriculture, Forestry, Fishing, Oceans, the Study of Biodiversity, Monitoring Urban Growth, Mapping of Wastelands and Managing Water Resources. These are of enduring importance, especially to the developing countries.

In the recently concluded World Conference on Science, COSTED organised a meeting on 'Science in Response to Basic Human Needs', the problems relating to the availability of adequate food and potable water to the people were thought of as potential areas of concentration as we step into the 21st century. Thus, it becomes necessary to understand the land use patterns, land cover changes, availability of recources such as water and use them in the most appropriate way to derive the maximum benefit. In these efforts, remote sensing techniques comes as a handy tool. In order to use this tool it becomes essential to educate all concerned on the effectiveness of this technology. It is heartening to note that this topic has been included in the school curricula in the developing countries in recent years.

Dr L R A Narayan, (formerly of India's National Remote Sensing Agency) a leading expert in remote sensing has been writing technical articles on remote sensing aspects in a very simple, easy-to-understand language suitable to the student community and to the general public. The present publication is a compilation of a series of such articles written by him for a leading Indian national daily, *The Hindu* over a period of six years.

COSTED—the committee on Science and Technology in Developing Countries—of the International Council of Scientific Unions has been publishing books, monographs and occasional papers on different aspects of science and technology for the benefit of the scientific community in the developing countries. We are pleased to make this publication available to students of science, young scientists and the wider public community everywhere, in the hope that it would help the learning process in the area of remote sensing, irrespective of individual backgrounds.

We gratefully acknowledge the financial support received from the Indian Space Research Organisation for bringing out this publication, and place on record our thanks to its Chairman Dr K Kasturirangan. The help rendered by

Dr S Krishnamurthy, Director, Publication Division, ISRO in the project materialising is specially acknowledged. Universities Press (India) Ltd., Hyderabad, have brought out this booklet in a very user friendly fashion and we also appreciate their timely publication.

September 1999

G THYAGARAJAN
Scientific Secretary
COSTED

List of Plates

Between pages 30 *and* 31

1

The Eye in the Sky

The concept of remote sensing from space platforms or satellites is too complex for the layman to grasp. Remote sensors have the powerful capacity to provide information about our Earth's surfaces. This first of a series of articles goes into the technical and related aspects of remote sensing.

There is a misapprehension in the minds of many, that data or pictures obtained from satellite altitudes are subject to security restrictions and are not easily accessible. Before going on to explain the basic principles of remote sensing from satellite altitudes and also bringing out its powerful capacity to provide enormous information about our Earth and activities taking place over it, I would like to re-emphasise here that remotely sensed data or pictures (imagery as it is called) is obtainable by anyone, covering any part of the country or even other countries, provided one is willing to pay for it.

Information obtainable from such data is enormous and can be used by anyone who needs information about our Earth's surface, including those of strategic importance.

"Remote Sensing" can be defined as obtaining information about an object by observing it from a distance and without coming into actual contact with it. In fact when we see an object and understand what it is, our eye is sensing that object remotely. This is a broad definition, but we generally use this term for observing our Earth's surface from space using satellites or from the air using aircraft which have been modified suitably.

Space platforms or satellite systems comprise

a) space segments

b) sensor systems and

c) ground segments.

India is one of the small list of countries which has all these segments available and operational.

The space segment consists of the satellite which could be put into orbit and in which a sensor system is mounted which, in turn, is the "eye" or the instrument which observes, in preplanned formats and transmits information to the Earth receiving stations. In some cases provision exists in the form of a tape recorder, which records the observed information, and transmits it to the Earth receiving station located at a different place other than the place which is being observed. Factually, the process of observing and transmitting to the Earth receiving station is more or less a continuous process. For example, the Indian Earth Receiving Station at Shadnagar, about 55 km from Hyderabad, has an antenna of 10 m diameter, with a 2 500 km radius visibility range. As the satellite appears in the horizon and comes within the observational range of the receiving antenna, observed data is transmitted which in turn is recorded and stored for further processing, till the satellite disappears from the visibility range. This is done automatically. When the satellite appears, the receiving antenna gets locked to the satellite and tracks it till it goes out of its visible range, like the Sun or the Moon rising and setting. With this range of visibility we are able to obtain data pertaining to India, and of adjoining countries like Pakistan, Afghanistan, Bangladesh and Sri Lanka.

The satellite, (for example the Remote Sensing Satellite — IRS-1A or 1B) is launched to an altitude of about 904 km, and orbits the earth in a north-south direction, and inclined by about 99 degrees. Each rotation takes 103 minutes in the case of IRS–1A & 1B, and may vary slightly in other satellites depending upon the design of the mission. While the satellite orbits around the Earth in a N-S direction (of descending node), the Earth itself rotates around its polar axis whereby the satellite observes different parts of the earth at each rotation, and the shift in the case of say the Indian remote sensing satellite is 2 872 km at the equator.

In other words, if the first satellite pass is over the eastern part of India, the next pass would probably be over the western part of India separated by 2 872 km. Similarly, the next day, the first pass visible to our Earth station will be 130 (approximately) km west of the previous days pass in the eastern part of India, and the second pass may go further west, beyond the visible range of our Earth station. Thus, we can see that our Earth station will be able to receive one or two passes every day, and it will come to the same place again after 22 days in the case of the IRS satellite, by which time it would have completed 307 orbits. The satellite revisit to the same place may be different for the various satellites, depending on the design of the mission.

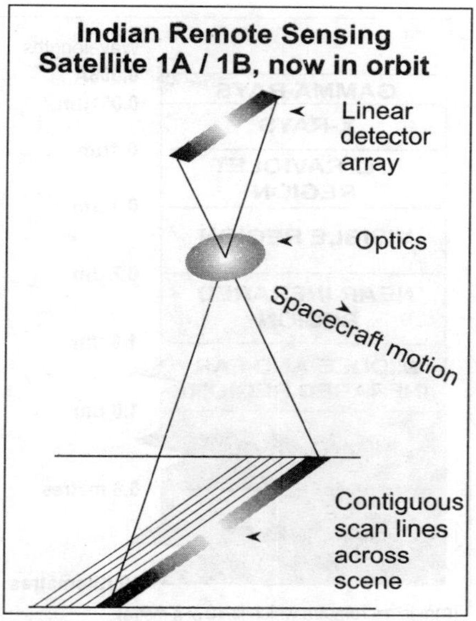

Fig. 1.1 *An illustration of the Pushbroom scan technique*

For example, in the case of the American satellite, Landsat 4 or 5, it is once in 16 days and in the case of the French satellite called the "SPOT" it is once in 26 days. The IRS satellite will cross the equator every day at 10.25 a.m. as per design, to give uniform illumination conditions.

Sensors are instruments which sense the objects on the surface of the Earth and records them. A camera is a form of sensor, which records the object it looks at and the light reflected reacts with the sensitised coating on the film which, when processed, reveals the object. Satellite mounted sensors, (they are called scanners) sense the various objects in the form of levels of grey in digital format of the energy reflected.

In the case of the satellite remote sensing generally prevalent the Sun's energy falls on to the object, and the part that is reflected is sensed by the scanners carried by the satellites, in different wavelengths. Each object on the surface of the Earth has distinct spectral reflectance characteristics called spectral signatures. The scanners keep moving from one side to the other across the line of flight or direction in which the satellite is moving. This way the scanner records continuously within its range. This varies in different satellites. In the case of the Indian remote sensing satellite, the scanner called LISS I, (Linear Imaging Self Scan) scans about 148 km of the surface of the Earth in each scene on a continuous basis and transmits them to the Earth receiving station. The spatial ground resolution obtainable in

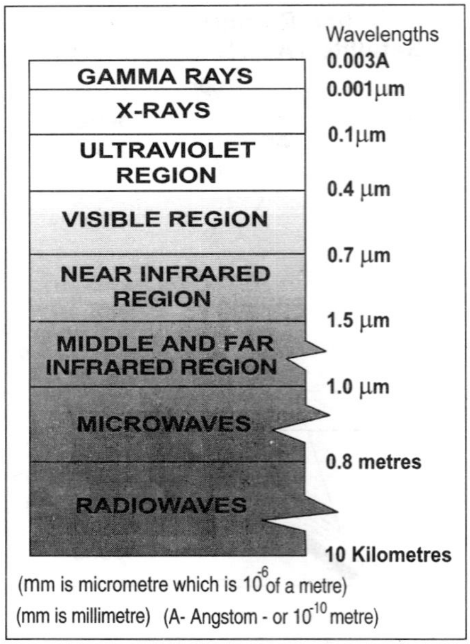

Fig. 1.2 *Different regions and corresponding wavelengths of the Electromagnetic (EM) Spectrum*

the case of the Indian remote sensing satellite, LISS I, camera is about 72.5 m on the ground. The data obtained in digital form can be converted to picture format by electro-optical processing. Thus, we get a picture or imagery of each scene covering on the Earth's surface an area of 145 x 145 km in the case of the IRS satellites (Plate 1.1). In the case of other satellites the area covered may be different. In the case of IRS there are two types of sensors, LISS I and LISS II. Each LISS II camera images (there are two of them) has a ground resolution of 36.25 metres on the ground and covers a swath of 74 km with a small overlap. Different satellite systems may have different types of sensors, as its payload, depending upon the design and based on specific needs.

Fig. 1. 1 illustrates how the Pushbroom scanning techniques are used for scanning the Earth's surface in the case of the IRS satellites, now in orbit.

Electromagnetic (EM) Spectrum

Remote sensing technology uses the visible, infrared and microwave regions of the solar radiation to collect information about the various objects on the Earth's surface. The electromagnetic spectrum has a visible region, which we use to view an object with our eyes. The Sun's energy, when it falls on the

Earth, gets partially absorbed and partially reflected. It is the part that is reflected that we generally use in remote sensing.

The different regions and corresponding wavelengths in the case of the Electromagnetic spectrum are illustrated in Fig. 1.2.

The reflected part of the visible region is only between 0.4 mm and 0.7 mm, as such there are many parts invisible to the human eye. In the case of remote sensing we also use infrared and microwave region by suitable devises, which will be explained late on.

Different objects reflect in different parts of the electromagnetic region, exhibiting their own spectral signatures. (See Fig. 1.3)

We can thus conclude that different objects manifest themselves predominantly in certain wavelengths of the electromagnetic spectrum in the visible range within certain specific wavelength ranges.

As illustrated in Fig. 1.2, within the visible region, we have different wavelengths within which colours from violet to red are exhibited, (the rainbow colours). When all these colours merge we see it as white light. If we therefore use filters and have sensors active in certain regions of the spectral range of colours, we can get information in different bands of the electromagnetic spectrum, as illustrated in Fig. 1.4.

Fig. 1.4 shows the band separation approximately provided for in our Indian remote sensing satellite 1A/1B. The data is obtained in separate channels. We could however combine any three channels to obtain what we call a colour composite. The above channels or bands were selected to highlight Agriculture, Forestry, Geology and Hydrology.

Different satellite remote sensed data may have different bands and channels, which may include near infrared as well as thermal infrared. However, it was also considered that the above selected bands will be able to cater to most of the application areas under proper lighting conditions and cloud-free situations.

Band	*Spectral Range*	*Application areas*
1.	0.45 – 0.52	Coastal environmental studies (coastal morphology and sedimentation studies), soil/vegetation differentiation, coniferous/deciduous vegetation discrimination
2.	0.52 – 0.59	Vegetation vigour, rock/soil discrimination, turbidity and bathymetry in shallow waters
3.	0.62 – 0.68	Strong chlorophyll absorption leading to discrimination of plant species
4.	0.77 – 0.86	Delineation of water features. Land forms/geomorphic studies.

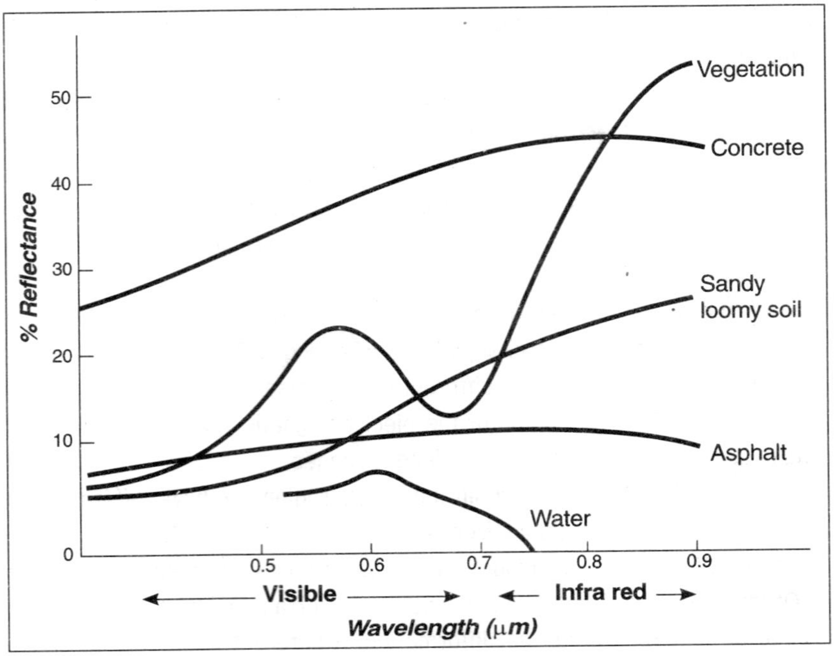

Fig. 1.3 *Spectral reflectance curves for selected features*

It was concluded in the case of IRS 1A/1B, after investigations, that the above mentioned application areas can be catered to against each of the bands indicated below (See Fig. 1.4):

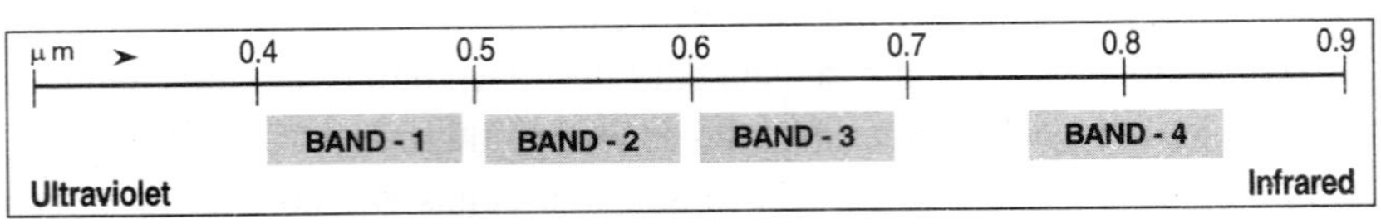

Fig. 1.4 *Band separation in Indian Remote Sensing Satellite 1-A/1-B*

Advantages of Satellite Remote Sensing

i) Synoptic view—A large or wide area can be covered by a single image/photo. Different satellite data with different types of sensor systems may cover different extent of areas.

ii) Repetivity—Can get the data of any area repeatedly at regular intervals of time, enabling monitoring of changes. Different satellites may have different revisit periods, according to the design.

iii) Coverage of inaccessible or difficult terrain like mountains, swampy areas, thick forests, etc. are imaged.

iv) Since data is obtained in digital form and in different channels, computer processing and analysis becomes possible.

v) Economic in cost and time.

Limitations of Satellite Remote Sensing

i) The general limitation in most cases is that when there is cloud cover, no data is obtained as the energy system in the visible ranges from Sun's energy cannot penetrate the clouds.

ii) Ground resolution is generally inadequate for detailed studies at the microlevel.

iii) In most cases, positional accuracy is not adequate for regular mapping activities with precision and geographical accuracy.

iv) Stereo viewing or three-dimensional viewing generally not possible, except in the French SPOT satellite data as of now.

Of late, the above limitations are being overcome by using sensors giving data in the microwave region of the EM spectrum, which can penetrate the clouds with its own energy systems. Improved and high resolution sensor systems have been developed, to obtain ground resolution of even up to 5 m on an average. Positional accuracies have also been improved and by using ground controls or satellite survey positional control systems like global positioning systems a much higher degree of accuracy has been achieved.

2

Keeping Track of the Satellite

This article focusses on the role of the ground segment to monitor, control and track the satellite paths and process the data beamed by it.

In all satellite operations, there is always a ground segment, which becomes operative as soon as the satellites are launched. Such a segment is generally situated in different locations, in touch with the satellite at all times, controlling its movements and orbits, and performing various other functions. For example, in the case of the Indian Remote Sensing Satellite, the major elements of IRS-1A ground segments are:

i) Telemetry, tracking and command (TTC) network

ii) Space control centre (SCC)

iii) Data reception system

iv) Data products system

The data reception and data products system receive the remotely sensed data signals, process and distribute them to various users in different formats and levels. The various levels of processing are Level-0, Level-1, Level-2, Level-3 and Level-4.

At each of these levels of operation, different corrections, such as the Earth's rotation and Radiometric corrections, Geometric corrections and precision corrections using ground control prints, etc. are carried out. In the Level-4 processing, certain enhancements of the images are made to bring out special features that aid in the interpretation. The ultimate products will include digital data, black and white transparencies, and colour infrared composites in transparent form, all of the format size of 240 mm. Hard copy prints can be made out of these in enlarged formats also. Similar activities are carried out in the case of all remote sensing satellites.

Similarly, for all other satellites, it becomes essential to have full fledged ground segments to monitor, control and track the satellite paths.

When a remote sensing satellite beams the data it has gathered, it is captured by the Earth receiving station and processed to enable the data to be converted in to a usable form and also apply various corrections. Either the data can be used in digital form using appropriate computers, or it can be processed using electro-optical instruments to produce hard copy picture prints in black and white or in colour or false colour infrared imagery form. Such imagery can be enlarged to relevant scales for interpretation and use by the application scientist to develop different themes and produce thematic maps.

Such data will represent the area of observation of the surface of the Earth as it existed at the time of the satellite pass. Since the satellite will pass over the same area again after some interval of time (say 22 days in the case of IRS), we can see what changes have taken place in the intervening period. Thus, we achieve a temporal resolution in each case, depending upon the revisit period, besides spatial and spectral resolution (variation in spectral bands), which are all fixed in the case of each remote sensing satellite system.

Types of Satellite Remote Sensed Data Available and their Characteristics

When the first satellite called "Earth Resources Technology Satellite" (ERTS) was launched by the US in 1972 (which was later on renamed as 'Landsat' 1, Land Observation Satellite), it was considered experimental, as there was no experience as to how such data be used, no appropriate techniques developed for their application, no adequate institutional infrastructure and no suitable equipment in either industrialised or in developing countries. However, the significance of such data in the natural resources sector was quickly realised as an unprecedented potential data source for various applications, particularly environmental monitoring and resources assessment. The sensors used in Landsat I and subsequently Landsat II, provided a four band multi-spectral scanner (MSS) operating in green (0.5–0.6 mm), red (0.6–0.7 mm) and near infrared regions (0.7–0.8 mm and 0.8–1.1 mm). Landsat I and II also carried three differently filtered Return Beam Vidicon, (RBV) cameras operating in green (0.475–0.575 mm) red (0.58–0.68 mm) and near infrared (0.69–0.83 mm) with an instantaneous field of view (IFOV) or resolution of about 70 m and a swath of 185 km of the ground.

The Return Beam Vidicon system was later abandoned as it was not found useful. The Landsat III, carried an MSS with an additional band in the thermal infrared region (10.5–12.5 mm).

Later, a 7-band mechanical scanner as a part of the Landsat 4 and 5 payloads was launched in July 1982 and March 1984 respectively, with a ground resolution of about 30 m except in the case of the thermal band 6, which had a 120 m resolution. The swath was 185 km of the ground, but it was orbiting at a lower altitude of 705 km compared to 925 km in the case of Landsat I, II and III.

This 7- band sensor was named the Thematic Mapper (TM).

The TM provided data in 7 spectral bands of 0.45–0.52 mm, 0.52–0.60 mm, 0.63–0.69 mm, 0.76–0.90 mm, 1.55–1.75 mm, 2.08–2.35 mm and 10.4–12.5 mm. This was a big breakthrough from the Landsat I, II and III systems. While the Landsat I, II and III had an 18 day revisit cycle, the Landsat IV and V Thermatic Mapper had a 16 day revisit cycle and when both were in operation, we were able to get data once in every eight days.

High Resolution visible (HRV) of SPOT

In 1986, the French Government launched a unique satellite called 'SPOT' with sensors with a charge couple device (CCD) and a pushbroom scanner, operating in three spectral bands, 0.50–0.59, 0.61–0.68, and 0.79–0.89 mm with a ground resolution of 20 m in multispectral mode and 10 m resolution panchromatic mode (black and white), covering a 60 km swath on the Earth. The most interesting feature of this scanner is its ability to provide stereocoverage by pointing the mirror to 27 degrees cross-track. This mirror steering capability also helps in providing a 5-day revisit cycle, although without the steering of the mirror pointing towards the nadir, it provided a 26-day revisit cycle. Spot 2 which was launched later on had similar sensors on board. Thus, the SPOT satellite was a quantum jump from the previous satellite system, although it had only 3 spectral bands in the multi-spectral mode compared to the Landsat-TM, etc.(Plate 2.1)

The Indian Remote Sensing Satellite - (IRS I & II)

As explained earlier, the India Remote Satellite (IRS-1A) was launched on March 17, 1988, with a 22-day repeat cycle, and orbiting at an altitude of 904 km. It carried two camera systems called Linear Imaging Self Scanning Camera I and II (LISS-I & LISS-II). The LISS I camera provided a resolution of 73 m while the LISS II provided a resolution of 36.5 m. The spectral bands covered are 0.45–0.52 mm, 0.52–0.59 mm, 0.62–0.68 mm and 0.77–0.86 mm. The swath width covered by LISS I was 148 km on the surface of the Earth, while the LISS II (2 cameras) covered a swath of 74 km on the surface of the Earth. In this case also different types of data products have been made available to meet the needs of different users, both in digital form as well as transparency and hard copy, paper print form.

We have other satellites like the NOAA (National Oceanographic and Aeronautic Administration) of the US, carrying on board a sensor called the Advanced Very High Resolution Radiometer (AVHRR), covering a vast ground area of about 4 million sq km and orbiting more frequently every day. This is also called an environmental satellite or weather satellite, and is accessible for reception in India as well. Besides this, the Europeans recently launched a satellite called the European remote sensing satellite (ERS-1), carrying sensors in the microwave range of the electromagnetic spectrum, which can penetrate through clouds and is particularly useful when other remote sensing satellites are unable to gather ground information due to cloud-cover. It has a resolution capability of about 30 m. Data from this satellite is also accessible to us, regularly being received at the earth station near Hyderabad, by the National Remote Sensing Agency (NRSA).

Data from satellites launched by various countries are not received in India at present. However, now, data from Russian satellites with a very high ground resolution of nearly 5 m is available for purchase, either directly from Russia or from their agents elsewhere. The Russian reconnaissance satellite pictures of a resolution of 2 m are also available.

3

Applications of Remote Sensing Data

This article, dwells at length on the applications of data provided by these systems. Promising areas of applications include disasterassessment, drought and environmental monitoring, and geomorphological mapping.

There are three types of resolutions in remote sensing from satellites; spectral resolution, which characterises the spectral bands in which the satellite sensors function and give us the data; spatial resolution, which characterises the extent of area or dimensions at which the sensor is able to distinguish and give us the data; and the temporal resolution, which characterises the periodicity at which the satellite reappears at the same area, at specified intervals.

In the case of spectral resolution, for example, while the American Landsat 4 and 5 satellites which carry a thematic mapper as its sensor, provide us data at 7 bands or channels of the EM spectrum, the Indian Remote Sensing Satellite provides 4 bands or channels of the EM spectrum. Similarly, while the Landsat TM provides a spatial resolution of 30 m on the ground, the IRS satellites provide a ground resolution of 36.5 m and 73 m using two cameras/sensors. The French satellite SPOT provides a spatial resolution of 10 m and 20 m.

In the case of temporal resolution, the Landsat TM reappears at the same place every 16 days, while the IRS reappears once in 22 days and SPOT once in 26 days, under the normal mode of operation. With these prime characteristics, and depending upon specific needs, one can choose the type of data sets that would be required for a particular study.

In fact, there are other types of satellites catering to certain specific information generation, like the NOAA series of environmental satellites. With such a variety of data sets available on a routine basis, these have been processed to provide information on various application aspects in the area of natural resources assessment and management, which is of prime importance to developing countries.

Remote Sensing Applications

While the technological breakthrough in satellite based remote sensing gained momentum, the application aspects took some time to catch up. Since satellite remote sensed data was available in a digital format, soon the use of computers for analysis and interpretation took roots. As the data actually depicted the state of the land as it existed at the time of observation, it was like bringing the ground to the office in the form of a picture to be studied by the various specialists. However, even if interpretation can be done by experts fairly clearly from remote sensed data, a short visit to the field to verify the facts cannot be avoided, before a final view is arrived at.

In the initial stages, the information obtained was more qualitative. But with time it became quantitative and more reliable. This was also possible due to improved spectral and spatial resolution.

Now, data sets available from a variety of satellites have opened up tremendous possibilities of extracting a variety of information. In several cases of resources studies, the technique is operational, catering to several varied needs in the area of management of natural resources. The major advantage in using satellite remote sensed data is that it is the latest available at any give time.

In space technology and its applications, the development group wants to be aware of the user's needs—users try to understand the capabilities and constraints of these new tools and their evolutionary trends. Policy makers wish to know in clear terms what technology can do and at what price. Hence, it is in these contexts that we have to view the total system of remote sensing both from the advantages it offers and the limitations it has.

Evidence in India and elsewhere gathered in the last decade has shown that remote sensing data either from satellite or aircraft have a wide scope and capability for carrying out fast and repetitive inventory, mapping and monitoring of resources. Application potentials of remote sensing techniques on resources related to some of the various disciplines are described here.

i) Geology and geomorphology mapping: Geology has a long history of remote sensing applications and is useful in

a) preparing small-scale reconnaissance maps of unmapped, inaccessible areas
b) updating the existing geological maps
c) rapid preparation of lineament and tectonic maps
d) identifying features favourable for mineral localisation.

The geomorphological maps find their utility in planning, management and monitoring of several other resources aspects.

Satellite data has been used in

a) preparing regional geomorphological maps on a scale of 1:250 000.

b) through repetitive coverage, monitoring and change defection in "exclusive domains," like sand-dune movements in deserts, river, resources, coastal regions, etc. The geological and geomorphological maps provide a preliminary base for providing alignment of transportation and communication routes and sites for irrigation, energy and industrial projects. They have also been used for delineating mineral and groundwater potential areas.

ii) Soil classification mapping: Soil is a three-dimensional continuum which depends on parent material, geomorphic processes, and environmental characteristics. Remote sensing data can study surface manifestation, (geomorphic) improving the efficacy of soil surveys.

iii) Water Resources: Understanding water resources is important to agriculture, industry, commercial establishments and residential areas. Water supply to these areas depends upon the available resources, both in terms of quantity and quality. Remote sensing data is useful in

a) surface water and stream pattern mapping,

b) pre-flood and post-flood river configuration and flood plain mapping,

c) snow cover mapping and run off prediction,

d) irrigated area studies and its monitoring

e) determining potential ground water zones. (See Plate 3.1)

iv) Agriculture (crops): One basic information that remote sensing can provide to agriculture is data related to crop identification and area measurement under different types of crops, or acreage estimation. This enables us to somewhat estimate the total production by understanding the yield per unit area. Such information has far-reaching consequences in providing adequate food security.

v) Forestry and vegetation mapping: Remote sensing can aid in providing: a) information about the extent of forest cover and give a general idea of the types of forest cover b) detection of forest hazards like fire, disease and excessive felling.

vi) Land use mapping: The knowledge of spatial distribution of land cover/ land use of a large area is of great importance to regional planners and administrators. Conventional ground methods are time-consuming, expensive and not uniform. Remote sensing data can be used to great advantage in obtaining up-to-date land use pattern of large areas at any given time and also monitor changes that occur from time to time.(See Plate 3.2).

Remote sensing can also provide data related to ocean and coastal zones like identifying potential area of fish concentration, environmental degradation that takes place in coastal zones due to over exploitation, etc. Other promising areas of applications include, disaster assessment, drought monitoring, environmental monitoring, etc., all of which have advanced significantly.

4

Remote Sensing over Fishing Zones

Remote sensing methods have helped greatly in the optimisation of ocean resources. We can envisage a stage when people will even be willing to share the expenses for these services.

As 60 to 70 percent of the world's population live within 20–30 km of the coastline, coastal zone management and optimisation of ocean resources have grown in of importance. The study of the ocean encompasses its physical, chemical and biological properties and its interaction with land. The biological productivity and resulting marine resources are other important aspects. Ocean related study was difficult in the past owing to a lack of adequate data. With the advent of remote sensing methods using satellite and aerial survey, the data coverage and accessibility have increased. Several parameters relating to the oceans can he studied using satellite and aircraft data.

For example, the ocean colour as measured by a sensor is found to bear a direct relationship with the suspended material of the water, thermal characteristics and the location of greenish biological matter. One of the important parameters that can be measured with sufficient accuracy is the Sea Surface Temperature (SST), which has been related to the concentration of fish population. This is an example of how a high technology has been applied at the grass-root level increasing the earning capacity by increased fish catches.

While other ocean related applications have become possible on remotely sensed data from satellites, on a near real-time basis, here, we shall see how a particular satellite and the sensor mounted on it have been used to obtain sea surface temperature and map it, on a regular basis, and pass it on to the fishermen who could concentrate on high potential areas and improve the catch.

Data available from TIROS-N, launched by the NOAA (National Oceanic and Atmospheric Administration) of USA, carrying a sensor called AVHRR (Advanced Very High Resolution Radiometer), has been used in obtaining sea surface temperatures, mainly from the Thermal Channel it is equipped with (See Plate 4.1). The wave length of the four Channel data it provides are, Channel 1: 0.55–0.90 micrometer, Channel 2: 0.75–1.0 micrometer,

Channel 3: 10.5–11.5 micrometer (Terminal Channel) and Channel 4: 3.55–3.93 micrometer (Middle infrared). The spatial resolution of this data is approximately 1 km and covers a large area.

While India has a 7 500 km long coastline, the fish catch is rather low. While viewed against the world production of 90 million tonnes, India's contribution is estimated to be only about 2 percent. Remote sensing can provide a significant part of the information needed to assess and improve the potential yield of the fishing grounds and also assist judicious exploitation, conservation and management of marine resources. Timely forecasting of Potential Fishing Zone (PFZ) information can help in maximising the efforts and thus save on fuel and help in optimising the scheduling of mechanised fishing operations.

The major oceanographic features, i.e. temperature and colour of the ocean, influence short-term forecasting of fishing grounds. In the absence of ocean colour information, the experiments have been using the sea surface temperature (SST) derived from NOAA- AVHRR satellite, all over the world. The use of SST in locating fishing grounds is at different stages of development in different countries. SST serves as a very useful indicator of prevailing and changing environmental conditions and is one of the important parameters which decides suitable environmental conditions for fish aggregation.

A national programme on "Ocean Related Remote Sensing Applications" sponsored by the Department of Ocean Development, is now being executed by the National Remote Sensing Agency, Hyderabad, in collaboration with agencies like the Space Application Centre (ISRO) Ahmedabad; National Institute of Oceanography, Goa; CSIR centre for mathematical modelling and computer simulation, Bangalore; Central Marine Fisheries Research Institute, Cochin; Institute for Ocean Management, Anna University, Madras and the Orissa Remote sensing Application Centre, Bhubaneswar.

Under this programme, the SST over the Arabian Sea and the Bay of Bengal (See Plate 4.2) is being retrieved using the NOAA-II satellite data. Based on SST, thermal features such as eddies, current boundaries and thermal gradients are recognised and a forecast indicating the Potential Fishing Zone (PFZ) is being generated. Data on SSTs of both coasts of India are available in the form of maps as well as statistical data.

Preparation of Potential Fishing Zone Maps

SST images over three or four days are composited and the minimum and maximum temperatures are noted down. These values are processed to

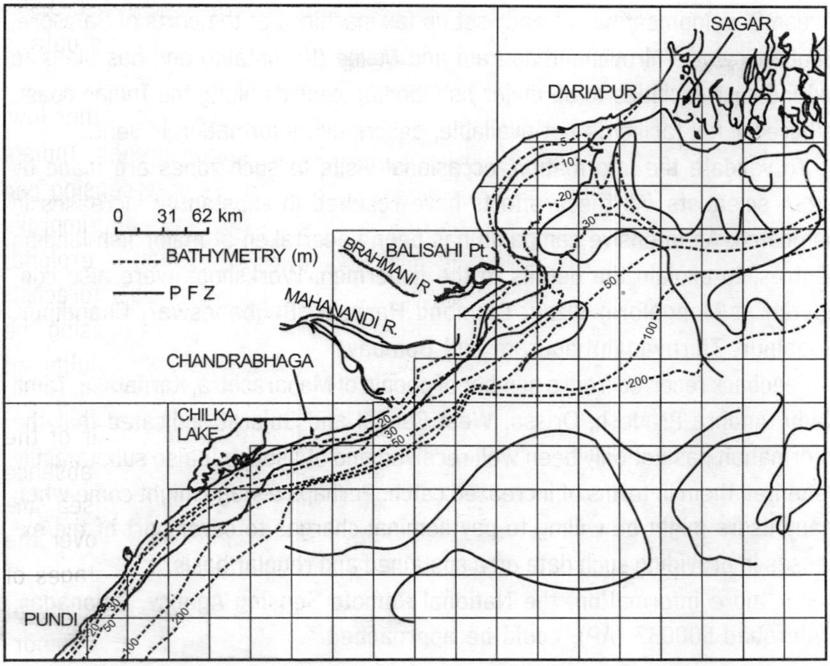

Fig. 4.1 *Potential fishing zones along the Orissa and West Bengal coast. (Source: NRSA)*

obtain maximum contrast of the thermal information. These are filmed to prepare relative thermal gradient images.

From these images, features such as thermal boundaries, relative temperature gradients to a level of 1 degree centigrade, level contour zones, eddies and upwelling zones are identified. These features are transferred using optical instruments to corresponding sectors of the coastal maps prepared with the help of Naval Hydrograph charts. Later, the location of the Potential Fishing Zone (PFZ) with reference to a particular fishing centre is drawn by identifying the nearest point of the thermal feature to that fishing centre. The information extracted consists of distance in kilometres, depth (for position fixing) in metres and bearing in degrees with reference to the North for a particular fishing centre.

The PFZ maps (Fig 4.1) thus prepared are sent through facsimile transmission (FAX) to major fishermen's associations, unions and Governmental organisations such as the Central Marine Fishery Research Institute (CMFRI), Fishery Survey of India (FSI), and state fisheries departments of all maritime states, including the union territories of Andaman and Nicobar and Lakshadweep Islands every Monday and Thursday. The Department of

Ocean Development has already set up fax machines at the ports of Balasore, Bhubaneswar, Thiruvananthapuram and Malpe (Karnataka) and has plans to extend such facilities to all major fish landing centres along the Indian coast. Wherever fax facility is not available, descriptive information is sent.

To validate the information, occasional visits to such zones are made by NRSA scientists. All these efforts have resulted in substantial increases in fish catch. An intensive campaign has been undertaken at major fish landing centres, to explain the details to the fishermen. Workshops were also conducted at Mangalore, Vizag, Diamond Harbour, Bhubaneswar, Chandipur, Gopalpur, Thiruvananthapuram and Bombay.

Feedback received from a number of people of Maharashtra, Karnataka, Tamil Nadu, Andhra Pradesh, Orissa, West Bengal and Gujarat, indicated that the information has not only been well received and utilised, but also substantially benefited them in terms of increased catch. Perhaps a stage might come when many users might be willing to pay nominal charges to cover part of the expenses in providing such data on a sustained and regular basis.

For more information, the National Remote Sensing Agency, Balanagar, Hyderabad-500037 (AP), could be approached.

5

Impact of Mining Activities on Environment-I

Remote sensing by satellite has become a valuable tool in monitoring environmental conditions. This article explains how iron ore mining has affected the environment over the years.

More and more people are becoming concerned about the environmental degradation taking place due to the so-called developmental activities or exploitation of land resources in various forms. One such important activity is mining to exploit mineral wealth, coal, etc. increasingly, most of the time not realising its impact on the environment on a long term basis. Till we had the availability of data from satellite altitudes, one could not assess the situation based on an overall perspective of a large area surrounding such activities, not only as a one-time measure, but also at regular intervals.

In areas of pollution, the result of uncontrolled development activities or exploitation of land resources could affect land, water and air, polluting them in one form or another.

With the availability of various types of satellite remote sensed data from 1972 onwards, it is now easy to understand the resultant degradation to our environment because of various human activities, and one such activity is mining (Plate 5.1).

Remote sensing from satellite altitude has emerged as a unique and extremely important tool in understanding and monitoring environmental conditions. Though aerial photography, the forerunner of present day satellite remote sensing has been in vogue for a very long time, the real spurt in the worldwide use of technology came with the launch of the Landsat series of satellites, the first of which was put in orbit in 1972, by NASA in the US The advantage of satellite remote sensing systems is that they provide a vantage point in space to study the Earth and its environment and provide a unique opportunity for inventorying, mapping and monitoring.

We now have several remote sensing satellites in orbit, including two of them called Indian remote sensing satellites (IRS-1A and 1B), beaming data to the ground at regular intervals, with improved characteristics. The simplest way of finding out changes over a period of time of any area, is to compare remotely sensed temporal data and demarcate the changed areas using simple visual or digital analysis techniques. The selection of the methods of analysis will depend upon certain considerations, the details of which are not elaborated here. Broadly speaking, remote sensing can be used for studying land, air and water, the three important facets of ecology and the environment.

This article is based on a study carried out by the scientists of the space application centre of ISRO, at Ahmedabad, covering the Kudremukh iron one mining area and the Dehra Dun-Mussoorie limestone mine belt, using remotely sensed data, from satellite altitude, and reported in 1990, as a project report funded by the Union Ministry of Environment and Forests.

Mining

A comparison of the 1951 and 1985 data indicates an increase of 42 percent in the total number of mines, which can be considered substantial. Likewise, the production of coal is reported to have increased 4 times, iron ore 10.65 times, copper 11 times, bauxite 33 times and limestone 16 times, during the same period. The result is we have been disturbing the ecosystem components and processes more and more and increasing the problems of environmental degradation. Some of the environmental problems normally associated with mining could be:

Water pollution due to wash on or overburden wastes, discharge of tailing into the water bodies, and mine drainage and coal-washing operations.

Air pollution because of release of noxious gases into the atmosphere and generation of dust during various operations.

Land degradation due to large-scale excavation, disposal of mine wastes and tailing and land subsidence (See Plate 5.2).

Landslips in hilly terrain caused by steepening of slopes during open-cast mining operations.

Deforestation due to felling of trees for construction of roads, and for fuel (See Plate 5.3).

Disruption of the water regime.

Problem of noise and vibration due to blasting and operation of heavy machinery.

Human–environment problems, such as settlements, health, etc.

Damage to sites of cultural, historical and scenic importance.

Problems caused by mine fires and degradation, air and thermal pollution, disruption of transportation, etc.

Methodology

The methodology adopted was to study and interpret the satellite remote sensed data of different periods of time, covering the same area, obtained at same season and registered with the help of the Survey of India 1:50 000 scale topographical map base. The analysis was carried out both by visual interpretation and by computer aided interpretation. The interpretation carried out in the office was spot-checked by personal visits to the area, to ensure correctness.

Kudremukh: Kudremukh is situated in the Aroli-Gangamula range of the Western Ghats in the Chickmagalur district of Karnataka. The river Bhadra flows through the area. This region was first surveyed in 1913, by the late Sir P Sampath Iyengar, a geologist of the Mysore Government (Chikka, Reddy, 1987). The resources lay unexploited until the Kudremukh Iron Ore Company started operating in 1977 in a big way by mechanised open-cast methods, mainly as an export venture.

The main environmental problems that resulted were:

Decrease in the forest area and degradation of 'shoal' forest in the valleys, around the mining area and residential complex.

Water pollution of the Bhadra due to spill-over from the tailing dam and run-off from the mining areas during the rainy season.

For the present study, satellite remote sensed data pertaining to the seasons October-December and February-May were used to identify and delineate various vegetational and other land use classes.

All other available maps and collateral data were also used in addition to ground visits. Satellite data of 1976 and 1985 were used for this comparative study.

The results showed:

The area under open forest 'shoals' has reduced by 26.7 percent, the main reason being submergence the reservoir.

The total grassland area has increased by 2.7 percent.

Mining activity has degraded the quality of 'shoal' forests around the Kachiga and Kudremukh rivulets, probably due to deposition of loose material.

The Bhadra river is highly polluted.

However, it was also noticed that the evergreen Western Ghats forests have not undergone any significant change either in extent or in density during this study period (See Plate 5.4).

The reliability of the study was estimated at approximately 90 percent.

Dehra Dun–Mussoorie Mine Belt: The forest around Dehra Dun conform to the Bhabar-Dun sal forest, while around Mussoorie it is lower western Himalayan temperate forest. The Dehra Dun–Mussoorie area has abundant water resources. Mining activity in Doon Valley started in 1911 and the number of mines kept increasing till 1985, when the Supreme Court ordered the closure of all but three mines under pressure from public litigation on environmental grounds (see Plate 5.1).

The main environmental impacts were:

Deforestation and degradation of forests

Shallowing of river/channel beds and changes in river courses

Increased incidence of landslips.

Drying up of Perennial Springs

Due to the non-availability of authentic data earlier, the study was based on satellite remote sensing data of 1972, 1985 and 1988 for analysis and to study the status of the environment in the area. Ground verification was also carried out. The methods followed were similar to the one followed in the case of the Kudremukh area.

The results showed that the forest cover has reduced by 2.8 sq km during 1972 to 1988. Notably, 3.28 ha of forest cover near Hatahipaon hill has altogether been obliterated due to intense limestone mining. The area under agricultural land has also shown a marginal reduction during the same period. Several perennial springs have dried up and water flow has reduced. Due to heavy landslips, some river beds have become shallow.

The above study illustrates how satellite remote sensing has been used to monitor environmental degradation that has taken place, specifically due to mining activities, in two locations.

6

Impact of Mining Activities on Environment-II

Data obtained from satellite remote sensing clearly shows problems arising out of uncontrolled exploitation of natural resources. Such indications could not have been observed but for the availability of space technology.

In this article I shall talk about the effects of uncontrolled mining activities on the environment and how it causes degradation of adjoining forests, pollutes river systems and, in some cases, affects the river systems and springs. While development processes are necessary, it may be prudent on the part of the authorities to carefully evaluate the deleterious consequences and try to evolve strategies to reduce such effects, if not totally eliminate them. We must attempt to protect our environment in all its aspects as best as we can, although it may not be possible cent percent.

In this article, we shall see how the setting up of large super thermal power stations affects the environment around the area, by analysing two typical examples, viz. the Korba super thermal power station in Madhya Pradesh and the Talcher super thermal power station in Orissa, using satellite remote sensing data. (see Plate 6.1)

A number of thermal power stations have come up in the Bilaspur district of Madhya Pradesh generating about 3 500 mega watts of power and, in the process, releasing about 45 tonnes of sulphur dioxide a day into the atmosphere. The first thermal power plant at Korba was commissioned in 1959. More thermal power stations are in the pipeline because coal is easily available from nearby Manikpur, Rajagmar, Kasumunda, Bankimogra, Gevera and Dipka and water from the Hasdo river. In view of the large-scale coal mining and power generation, the problems of environmental degradation have assumed critical dimensions.

Unless urgent remedial measures are taken the situation is sure to deteriorate. The main problems are:

Problems caused by flyash deposits and other gaseous pollutants.

Degradation of the quality of water from the Hasdo.

Deforestation and degradation of forests.

The main polluting agencies at Korba are the power plants of the Madhya Pradesh Electricity Board (MPEB), and the National Thermal Power Corporation (NTPC). The Bharat Aluminium Company (BALCO), a number of coal mines and other ancillary industries add to the problem substantially (See Plate 6.2).

Use of Satellite Remote Sensed Data

The first satellite remote sensed data became available in 1972. Therefore, we had the opportunity to get an overall synoptic view only after l972. For evaluating the situation, in this case, satellite data of 1972, 1985, 1986 and 1989 were used, covering about 17 years, to have an idea about the changes in the land cover and the impact of thermal power generation and mining activities. Satellite data obtained from Landsat I, and Landsat V (Multi-spectral and thematic mapper data of 30 m spatial resolution) and Indian remote sensing satellite I- A (LISS II of 36.5 m spatial resolution) were used in the form of both colour composite pictures as well as in digital format. Besides interpreting these data, other supplementary information such as meteorological data and data collected from limited ground visits were also integrated. A total area of about 804 sq Km, within a 16 km radius from the mid-point of various power plants at Korba was analysed.

Results of Investigations

The land transformation in this area can mainly be attributed to mining (Plate 6.3) and power generating activities. The impact of thermal power plants on the land has led to a decrease in forest area, and large-scale flyash deposits and coal mining has choked the streams in their environs.

Scientists of the Space Application Centre of ISRO say that the total forest area which was 258.60 sq km in 1972, has been reduced to 192.91 sq km in 1989, a 25.4 percent decrease. Plantations which are supposed to compensate for the loss of forest have only marginally increased (2.30 sq km in 1972 to 7.61 sq km in 1988). However, the area where the forests were close and thick were not much affected, perhaps due to some kind of preventive measures. Increased human habitation and farming had also led to the loss of some forest area.

The table below shows the areas in square kilometres of various landcover in the 16 km-radius zone.

In the case of wastelands, it may be noted that there has been an increase from 29.6 sq km in 1972 to 45.71 sq km in 1989, mainly attributable to intensive coal mining, rendering a large area barren.

Category	*1972*	*1985*	*1989*
Total forests and plantations	258.60	200.30	192.91
Agriculture	428.16	438.96	438.35
Wastelands	29.60	46.20	45.71
Mining areas	0.40	5.62	8.84
Settlements	29.40	53.78	57.59
Water bodies	18.40	17.20	18.66
Ash ponds	3.50	6.00	6.00
Rivers	36.50	36.50	36.50
Total study area	804.56	804.56	804.56

Source: Project Report. RSAM/SAC/RSAG/ ENVN/PR/08/90 from Space Application Centre-ISRO

Smoke from the power plants at Korba and from the older units are often the reason for poor visibility. The direction of the plume is governed by wind direction and speed. The disposal of flyash is another critical problem (Plate 6.4). Normally, flyash slurry is collected in ash ponds, but this can lead to water pollution if the disposal system is not foolproof. In 1980, the National Environmental Engineering Research Institute (NEERI) Nagpur, conducted a survey in Korba, and recommended urgent remedial action to tackle the water pollution problem.

The Korba area is highly polluted and this has affected the health of human beings and animals, not to mention the damage to vegetation and the ecological changes. High incidence of bronchitis, asthma and pulmonary emphysema has been reported. Cattle suffer from anthrocosts and dermatitis, the mortality rate being high. Sulphur dioxide pollution in the region has led to lower crop yield, especially of vegetables.

The present study has indicated a 90 percent reliability using satellite remote sensed data, without which such studies covering a large area in such a short time, would have been practically impossible.

Talcher in Orissa is fast emerging as a big source of coal and thermal power. The area spreads over 1814 sq km and has about 39.626 million tonnes of coal, of which about 5 207 million tonnes of reserves have been proved. Mining at Talcher started in 1921, coal-based thermal power generation in 1967 and, as of now, about 470 mW power is generated.

The main environmental problems in the area are:

i) Deforestation and degradation of forests

ii) Changes in the land cover

iii) Deposits of flyash and air pollution

iv) Deterioration in the water quality of the Brahmani river

To study this area, satellite remote sensing data of 1975, 1985 and 1988–89 were used, that along with other supplementary data, as in the Korba study.

The method followed was also similar to the Korba study; the area covered was 16 km radius from the mid-point of the main coal mining and thermal power plants. The results are tabulated below.

Area in sq km under various landcover classes within a 16 km radius zone

Sl. No.	*Category*	*1975*	*1985*	*1989*
1.	Total forests	87.54	64.96	57.69
2.	Wastelands	10.75	13.61	14.68
3.	Waterbodies	3.72	4.00	4.00
4.	Settlements	53.27	57.05	57.05
5.	Mining Areas	0.69	1.94	2.83
6.	Rivers	36.37	36.37	36.37
7.	Ash Ponds	–	–	1.07
8.	Agriculture	611.22	6216.63	630.87

(Source: Space Application Centre ISRO – Ahmedabad)

Smoke plume from the Talcher power station could be closely observed on Landsat 5, satellite remote sensed data of 1987; it was 6.25 km long and 0.5 to 1.25 km in width in the south east direction.

The total area affected by flyash deposits and coal mining has increased from 25 sq km in 1979 to 117 sq km in 1987. Activities related to mining, power generation, fertilizer production and aluminium production have polluted water resources. But data revealed that water pollution was not as acute a problem.

The pollutants have significantly affected residents, animals and crops, more or less in a similar way as in the case of Korba.

This study using satellite remote sensing data clearly shows the various environmental problems that have resulted due to the uncontrolled exploitation of natural resources, considering only the economic benefits but not the social and environmental impacts. Such clear indications could not have been observed but for the availability of space technology in providing satellite observational platforms as an eye-opener.

These studies were carried out by the scientists of the Space Application Centre of ISRO, Ahmedabad, to highlight some of the advantages of satellite remote sensing.

Plate 1.1 Sample satellite imagery from the Indian Remote Sensing satellite. (Courtesy NRSA)

Plate 2.1 SPOT (French satellite) data of a part of Madras.

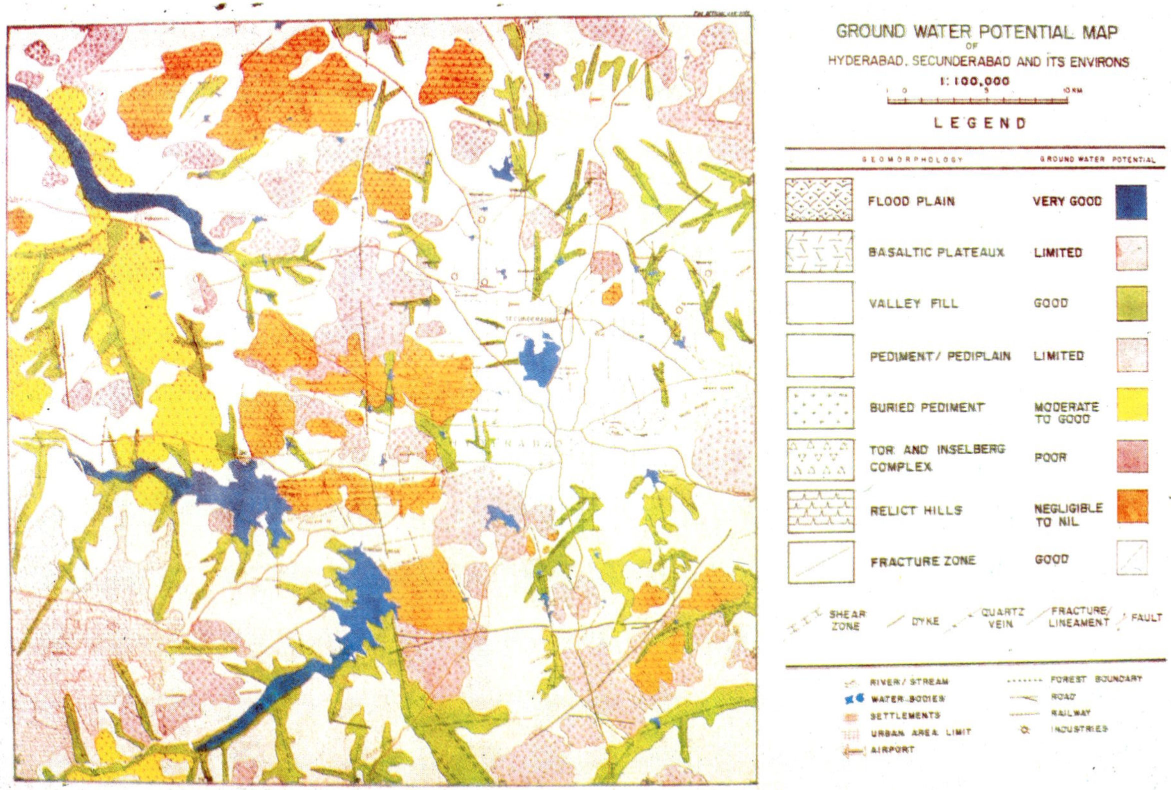

Plate 3.1 Potential groundwater zones in Hyderabad. (Courtesy AP State Remote Sensing Application Centre)

Plate 3.2 IRS-IB view of Allahabad (Courtesy NRSA)

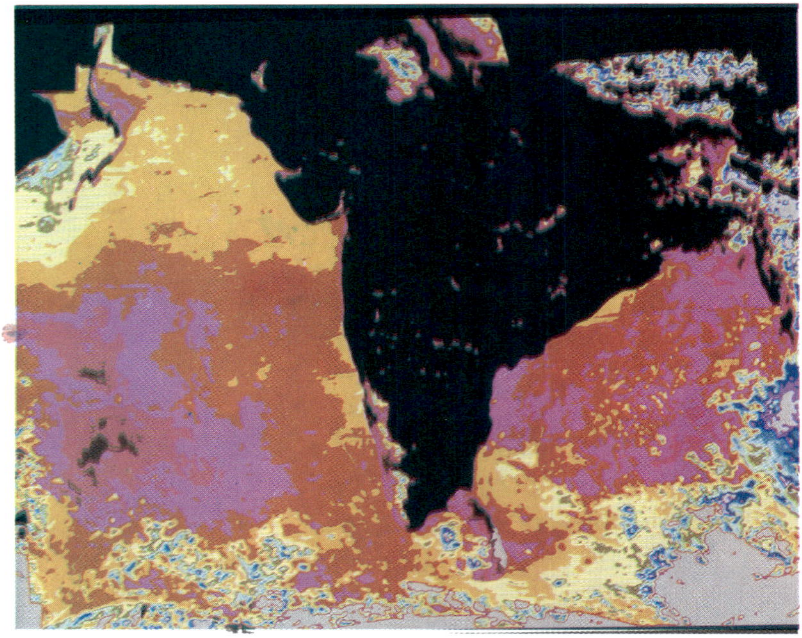

Plate 4.1 Sea surface temperature (SST) studies using NOAA-AVHRR Weather Satellite data. (Courtesy NRSA)

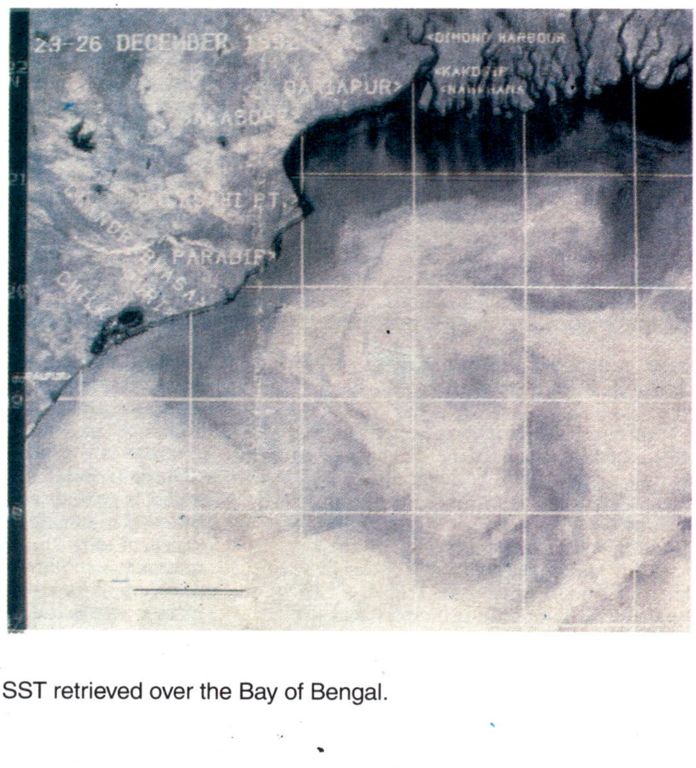

Plate 4.2 SST retrieved over the Bay of Bengal.

Plate 5.1 An abandoned mining area.

Plate 5.2 Tailing reservoir and grasslands in the background.

Plate 5.3 Decrease in the oak forest due to mining activity.

Plate 5.4 Closed evergreen western ghats forest.

Plate 6.1 NALCO captive power plant.

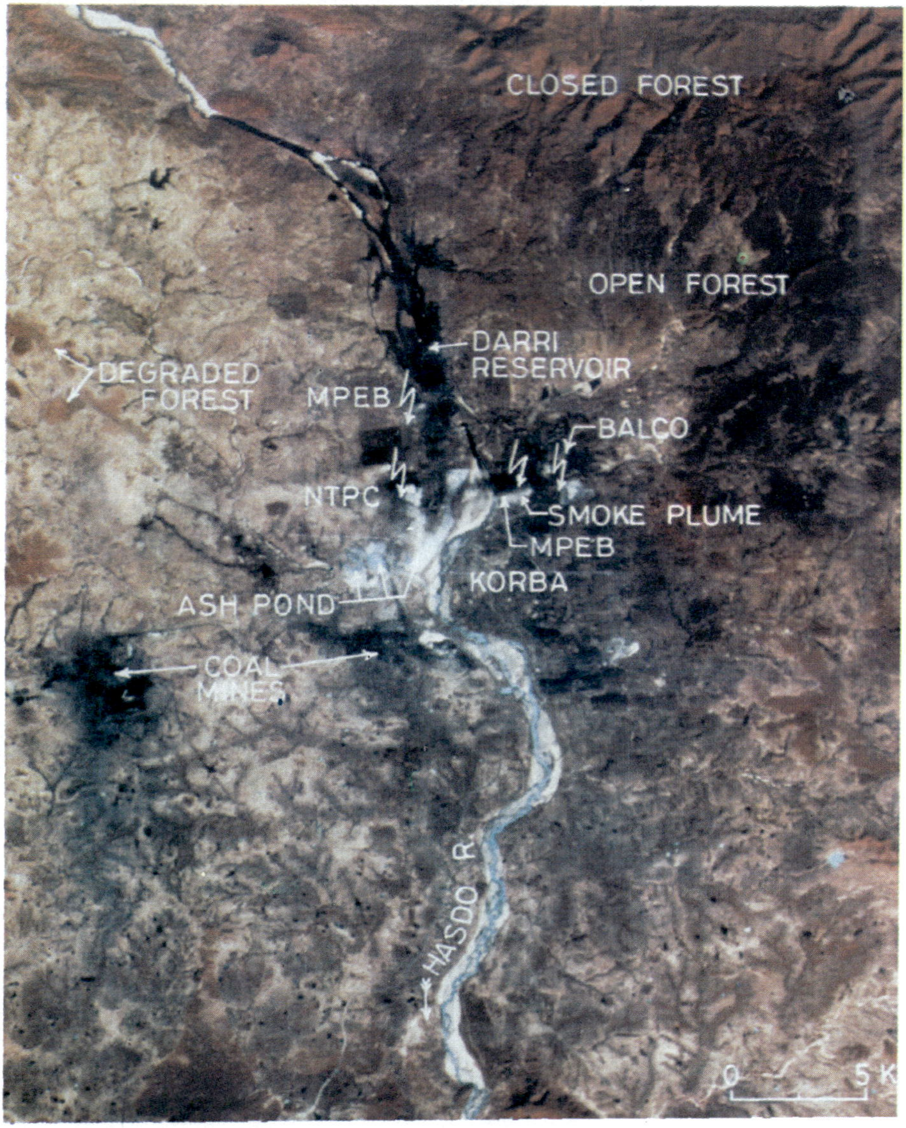

Plate 6.2 IRS-1A picture of Korba and environs.

Plate 6.3 Bharatpur open-cast mine.

Plate 6.4 The NTPC ash pond.

Plate 7.1 IRS satellite imagery of the north eastern region showing areas degraded by shifting cultivation in light blue; deep red indicates forest cover.

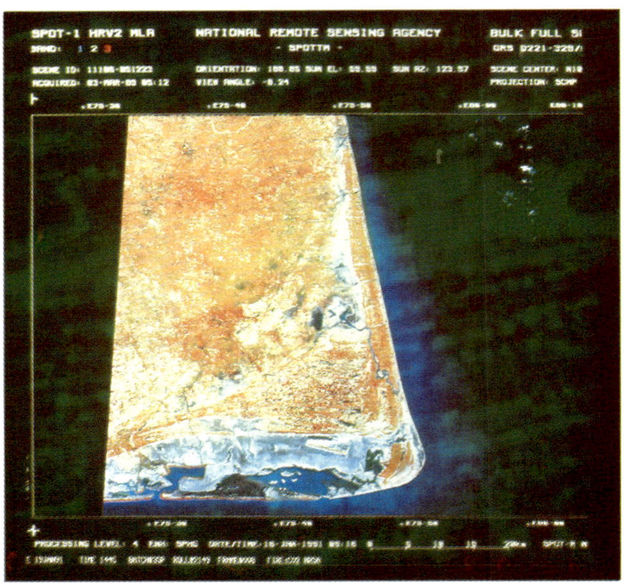

Plate 8.1 Satellite data showing various land use classes in the Vedaranyam area.

Plate 9.1 Shifting cultivation and smoke because of fire in Meghalaya. (Courtesy NRSA)

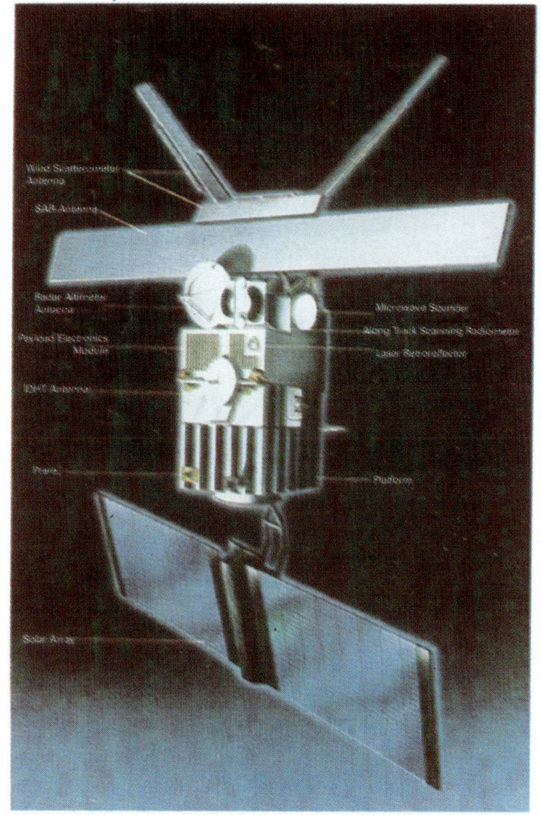

Plate 10.1 ERS-I Satellite and payloads

Plate 11.1 A tracking antenna which receives important data from a remote sensing satellite.

Plate 12.1 Ring road alignment around Bangalore. (Courtesy NNRMS-ISRO, Bangalore)

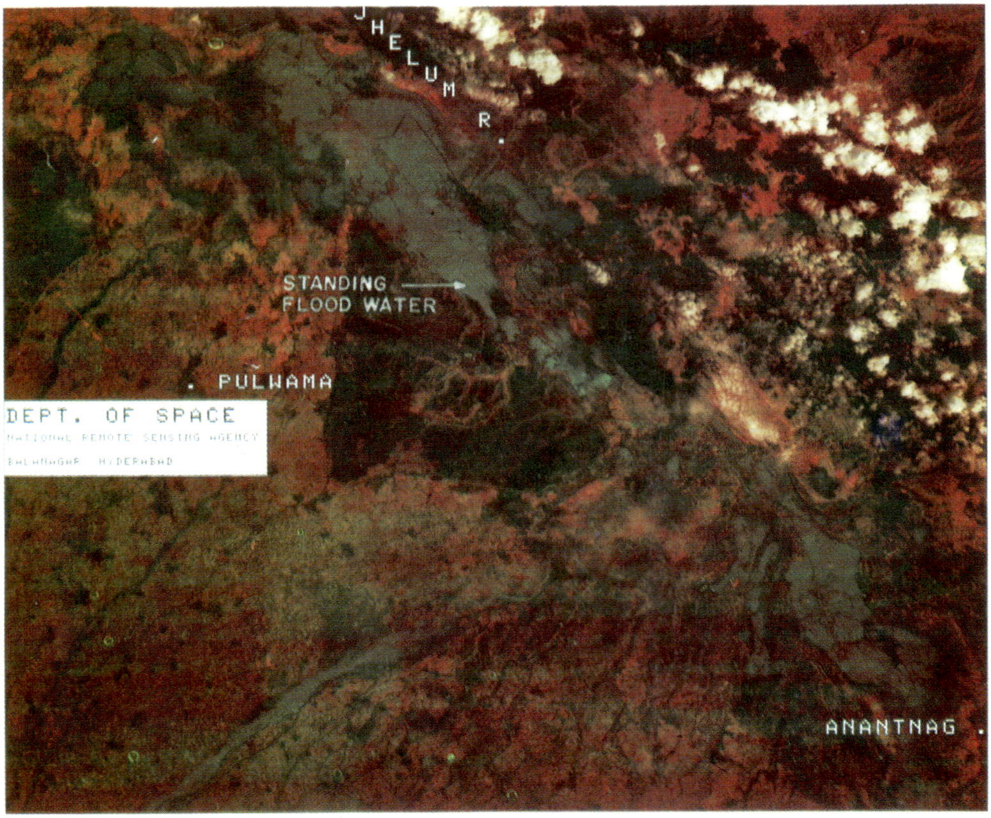

Plate 13.1 Satellite picture of Jammu and Kashmir before the floods. (Courtesy NRSA)

Plate 13.2 Satellite image showing the flood situation in Jammu and Kashmir on 13 September 1992.

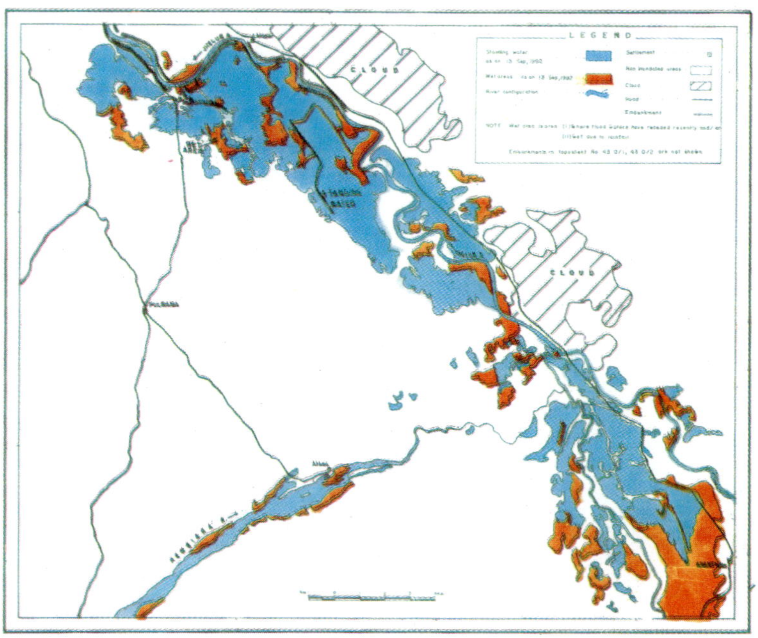

Plate 13.3 Flood map of inundated areas of Jammu and Kashmir.

(a) (b) (c)

Plate 15.1 SPOT data of Bombay metropolitan region: (a) Borivili (b) Thane (c) Kalyan

7
Remote Sensing and Biodiversity

Satellite based remote sensing lends itself to capturing details of the Earth's surface to understand the distribution and status of biodiversity to integrate human needs with the long-term maintenance of biodiversity on the Earth.

Biodiversity is most simply defined as the variety of life and its processes. It is the sum total of species richness, i.e. the number of species of plants, animals and micro-organisms occurring as individuals, populations, a community or an ecosystem. This means that biological diversity embraces all life forms—from plant and animal life to micro-organisms—and the water, land and air in which they live and interact. Biodiversity and the mark of human civilisation have one thing in common; they have some attributes that lend themselves to spatial display and analysis. As to its loss, it is recognised that there is no fixity of species, and extinctions and even corresponding renewals have been common in geological times.

While satellite based remote sensing lends itself to capturing details on the surface of the Earth, covering large areas in one scene (approximately 185×185 sq km) aerial platforms cover much smaller areas. Besides, the satellite based remote sensing facilitates more frequent observations, by periodic reappearance at the same place (for example once in 16 days to about once in 26 days by different remote sensing satellites in orbit). There is no such automatic facility in the case of aerial remote sensing, unless we make efforts to fly aircrafts at given intervals of time. However, the spatial resolution obtainable now in the case of aerial remote sensing (aerial photographs) is much higher than what we are able to get from satellite altitudes, although perhaps by the turn of the century, we may be able to get improved spatial resolution data from the satellite altitude also.

While only about 1.4 million different species have been known to us by description, it has been estimated that over 50 million species may exist on the Earth, if we take into account the diversity of invertebrates and micro-organisms. Most of these are found in the natural ecosystems of forests, savannas, pastures and rangelands, deserts, tundra, rivers, lakes and seas.

At the turn of the 20th century there were less than two billion people on Earth and biological diversity was mostly exhibited in museums. As we approach the 21st century, the global population is pushing six billion and biodiversity is causing great concern and is in the headlines.

The current decline in the world's biodiversity is largely the result of human activities, resulting in habitat destruction, over-harvesting, pollution, and the inappropriate introduction of foreign plants and animals. Urgent and decisive action is needed to conserve and maintain genes, species and ecosystems, with a view to the sustainable management of biological resources. Capacities for the assessment, study and systematic observation and evaluation of biodiversity need to be reinforced, for which newer technological means or tools may have to be used.

It is here that the use of remote sensing, particularly from satellite platforms, is becoming relevant and important. At large geographic scales—from watersheds to the entire biosphere—biological diversity includes variety in the kinds of ecosystems, their patterns and linkages across regional landscapes. It is stated in several places that one-quarter to one-half of all the species on Earth will disappear by the year 2100. "With their passing, humans will forever be denied the products and environmentally rejuvenating services those species once provided." Pessimists speculate on the ability of the planet to sustain human life itself, which may be seriously compromised. "We depend on other species for oxygen, food and medicines—what happens when they are gone?" (Blair Csuti)

A programme called "gap analysis" has been advocated by the US Fish and Wildlife Service by J Michael Scott of the Idaho Cooperative Fish and Wildlife Research Unit. It is essentially a biodiversity mapping programme. Its goal is to map the distribution of indicators of biodiversity, like plant communities and birds, to see which are already protected. Those that are not are the "gaps" that require conservation attention so that we can avoid all the costs and controversy attendant with turning them into endangered species.

Two technologies in the last quarter of the 20th century, Remote Sensing and Geographical Information Systems (GIS) have opened the door to understanding the distribution and status of biodiversity over larger areas. More than that, the patterns of current and future human land use can be mapped for the preparation of plans to integrate human needs with the long-term maintenance of biodiversity on Earth (See Plate 7.1). When ecosystems can be described in terms of dominant vegetation, as prevalent in North America and Australia, they lend themselves to mapping, using remote sensing technologies, as they make up the most visible part of an ecosystem. They also provide the structure that defines ecosystems—forests, shrublands, grasslands and so forth. Secondly, they integrate many of the other important features of an ecosystem such as temperature, rainfall, soil type, elevation, etc. In fact, gap analysis programmes use vegetation as a surrogate for ecosystems.

There are three fundamental data layers in gap analysis: vegetation, species distribution and the location of protected areas. Vegetation is mapped from satellite data such as the Landsat Thematic Mapper (TM) in digital form. Animal distribution is predicted by identifying their preferred habitat types within their distributional limits. Other items of information can be integrated with these depending upon the nature of their availability. In the US, the Idaho Department of Water Resources was the first to have a complete set of digital, thematic layers necessary to conduct gap analysis, using database layers of actual vegetation, management area boundaries, and generalised wetlands. Compilation of such large area data and its analysis is, of course, a major task needing appreciable monetary resources. To begin such a programme in our country, we may need co-ordinated efforts by several committed organisations with expert leadership which will certainly accelerate
the work being initiated under the conservation of biodiversity.

This article highlights the use of technologies like remote sensing and GIS as the topic of biodiversity has gained considerable importance throughout the world. Perhaps a programme like the gap analysis using advanced technologies in countries like ours may be highly beneficial and necessary to protect fast disappearing species of value to us, owing to increasing population and the effect of human activity upon various ecosystems. According to a publication of the world resources institute (Oct '89) between 1990 and 2020, species extinctions caused primarily by tropical deforestation many eliminate between 5 and 15 percent of the world's species. With roughly 50 million species on Earth, this would amount to a potential loss of 15 000 to 50 000 species per year or 50 to 100 species per day. Hence it calls for immediate action, to prevent serious losses. For this, there is an urgent need to have an up-to-date and near real time information system to enable sound decision-making. Remote Sensing can play a vital role in this effort.

8

Satellites and Coastal Zones

Mapping of coastal zones gives us insight about how to conserve its ecobalance, save the mangroves and implement effective coastal zone management.

Effective coastal zone management will need accurate and comprehensive scientific data, on which decisions can be based. Availability of satellite remote sensed data on a near real-time basis has proved to be a boon as it can provide a synoptic view on a repetitive basis. About 60 percent of the world's population live about 10 to 20 km along the coastline as mentioned earlier. Hence, coastal degradation processes need careful watch, which has now become possible by the regular use of satellite remote sensed data.

The satellite remote sensing enables us to locate and monitor tidal wetlands, coastal landforms, potential aquaculture sites, mangroves, estuary dynamics, shoreline changes and off-shore aspects like suspended sediment dynamics and coastal currents, near-shore bathymetry, internal waves, etc., besides monitoring pollution due to oil slicks.

Mapping of coastal wetlands, coastal landforms and shoreline changes provide considerable insight into various inputs to be taken into account for planning developmental activities in these areas. Before satellite remote sensing data became available, we could hardly see the effects due to various activities, as changes took place slowly and we could locate and map these only at a very slow pace. The information available was also highly localised and we could not comprehend inter-related effects of the various activities. For example, we did not know how setting up industries in certain areas would affect the coastal environment and upset its ecobalance.

Space application scientists made a rapid study of the coastal regions using Indian remote sensing satellite data, on a scale of 1:250 000 about four years ago and their study in brief revealed:

i) The South Gujarat coast experienced a lot of erosion in several places.

ii) The Karnataka coast is marked by spectacular spits.

iii) On the Tamil Nadu coast, erosion was noticed north of Madras, while depositional processes prevail at Point Calimere and Nagapattinam.

iv) In Andhra Pradesh, tidal flats and mangroves are reducing in extent because of reclamation processes. The Pulicat lake appears to be shrinking.

v) On the Orissa coast, mangroves appear to have been destroyed and removed near Karanjamal and Paradis port. The area of the Chilka lagoon has shrunk from 871 sq km to 790 sq km between 1973 and 1986. This, perhaps is mainly due to sedimentation in the lake. The scientists of the Space Application Centre (ISRO), Ahmedabad are now engaged in mapping wetlands on a scale of 1:50 000, in Kerala and Tamil Nadu. This will be useful for planning conservation measures.

Mangroves form an important vegetation belt along the coastal regions and its presence has considerable influence in maintaining the proper environment balance. Destruction of mangroves will, in the long run, have a serious impact on the coastal ecosystem. Mangroves to some extent protect land from cyclones and its ill-effects on crops, cattle and human habitation. This was clearly noticed in the case of two cyclones along coastal Andhra Pradesh, one when mangroves were there in fair measure and the next when considerable destruction of mangroves had taken place, by which flooding took place deep into the land. Observing such a phenomenon could not have been easily possible without resorting to the use of satellite remote sensing data.

Mangroves are considered a bridge between terrestrial aquatic ecosystems, and protect sustainable fisheries. Tourism and aquaculture threatening the existence of mangroves, in several places. The National Mangrove Committee in the Department of Environment tries to get an accurate extent of mangroves and supports measures to protect the coastal ecosystems from being degraded. Several activities cause reduction in mangroves, such as deforestation to provide fuel-wood, waste disposal in mangrove areas, mining activities, oil spills, active land reclamation for housing, ports, industries, roads, canals, salt pans and extension of fish ponds.

A study by scientists of the NRSA, Hyderabad, using satellite remote sensed data of 1973, 1985, 1990 and 1992, covering Andhra Pradesh ecosystems to delineate many of the cultivated areas and mangroves, revealed that in 1973, the area under prawn cultivation was almost negligible and by 1985 and 1990, it became more than five times in Guntur and 10 times in Krishna district. In this study, one could also notice the decrease in the area of mangrove vegetation suggesting that the increase in prawn cultivation may have to some extent, aftected mangroves. Mangroves, essential for maintaining ecological balance in the coastal ecosystems, are being destroyed in some of the areas as in Kottapalem in Repalle Mandal in Guntur district, for extending prawn culture. Remote sensing data not only help in the estimation of the area under brackish water aquaculture along the whole Indian coast, but can also help in selecting the suitable locations for prawn/shrimp farming without a serious threat to mangrove systems which are essential in the coastal areas. In Krishna district the prawn cultivation has increased from 1973 to 1992, while the mangroves have reduced during the same period as could be observed from satellite remote sensed data of this period. In terms of statistics, prawn cultivation was nil while the area under mangroves was about 5884 ha in Krishna district in 1973. In 1992, the area under prawn cultivation increased to 6005 ha, while the area under mangroves decreased to 5479 ha.

Some time ago, a similar study, but more comprehensive in nature was carried out by the Institute of Remote Sensing, Anna University, covering the Vedaranyam area, funded by the Ministry of Environment and Forests (See Plate 8.1). The objective was to inventorise the mangrove vegetation, the ecological status of forest in the coastal zone, map the wetlands in the area with emphasis on erosion and deposition trends and to study the impact of salt industries in the area. They appear to have used aerial photography of the 1976-77 period and later compared the results with satellite remote sensing thematic mapper data of 1986, i.e. leaving a gap of 10 years. Subsequently, the use of the French Satellite SPOT data of 1989 was also used and compared. The results showed reduction of mangrove forest by 40 percent, increase in salt pans to the extent of around 32 percent, perhaps at the expense of mangrove forest. In some areas, considerable marsh vegetation have been degraded. All these indicate rapid deterioration of the coastal environment due to excessive human activities.

It can thus be concluded that satellite remote sensing data has provided us with unbiased and near real-time data, which can be used to map and monitor coastal resources.

9 Monitoring Forest Cover

Gathering of real-time forestry information enables us to compile gross forest vegetation resources data of the entire country in a single format and monitor the changes.

So far we have discussed the various aspects of remote sensing technology and its applications, brought out the capacity of satellite borne sensors/ cameras to observe very large surface areas of the Earth and record them for us to study. Since remote sensing satellites reappear and re-observe the same areas at periodic intervals, we have the added advantage of monitoring changes that occur in the area of coverage, to plan for remedial measures for any adverse happening, much sooner than what we were able to do in the past, when the facility of remote sensing from space altitudes was not available. It is in this context that we could view forest cover or vegetation cover of a large area, and the changes that occur to it, as the forest is a renewable natural resource which is very vital for human survival. For balanced environmental sustenance, a 33 percent vegetal cover of the total land area is considered most desirable, from the point of view of a 'holistic' concept.

The forest area reported for 1973-74 (reported by Annon in his country report in 1978), was 74 964 million ha out of a total geographical area of the country of 328 779 million ha, which amounted to 22.80 percent and the per capita forest works out to approximately 0.13 ha. Although area-wise, the forest land was about 23 percent as per the Central Forestry Commission (1974), the productive forest cover in India by 1974 was only about 35 million ha (M Unni, NRSA).

Since there was no effective and objective mechanism to gather near real-time forestry information, the figures quoted by various people differed, and it was difficult to either agree with or to deny such statistics, authoritatively. Because of a steady increase in population, the demand on forest produce was increasing, and this resulted in gradual reduction in the forest wealth as a whole, although it was not immediately noticeable.

The forests being under the State's jurisdiction, the Centre depended on State officials for information, which, to some extent, was subjective, and could not be called consistent. With the launching of the first Landsat series of satellites with a multi-spectral scanner on board in 1972, we could get remote sensed data of our entire country, viewed and recorded in a near real-time mode.

In the meantime, development of interpretation techniques for different land-based natural resources was being developed, both by visual means and by using computers. The first three series of Landsat satellites had on board a multi-spectral scanner, giving a ground resolution of approximately 80 m and in four spectral bands. They also carried on board a sensor called Return Beam Vidicon Camera, the data from which was not made use of or favoured by many.

The original Landsat multi-spectral scanner imagery was on 1:1 million scale and it was possible to get these enlarged to a 1:250 000 scale for visual interpretation, using simple optical aids. Scientists were, in the meanwhile attempting to operationalise interpretation techniques to provide information on various natural resources, to cover large areas at a time, taking the advantages that the satellite remote sensed data provided, although not on a detailed scale.

By the early Eighties, remote sensing scientists representing various disciplines, came forward to undertake operational investigation projects using satellite remote sensed data products and achieved results of high dependability, which could be used for many practical purposes. Scientists also began to delineate forest cover types, and to conceptualise the ecological status of the forest.

The national remote sensing agency, established in 1975, took up a study to establish the credibility of using satellite remote sensing for forest surveys – to compile gross forest vegetation resources information of the entire country in a single format and monitor the changes. The idea of study was conceived by the then chairman of the department of space, Prof Satish Dhawan and carried out by the scientists of NRSA at Hyderabad. Essentially, it consisted of making a rapid assessment of the forest/vegetation cover as it existed in India in 1972-75 and again in 1980–82, to demonstrate the deteriorating condition of the forests and also to use remote sensing technology for making rapid assessments (Plate 9.1).

The result proved to be an eye-opener as it focussed on immediate attention to be given to the forestry sector as a whole, to improve our ecological stability. There were certain limitations and controversies but, considering the overall achievement, it was a great success, and this paved the way for the adopting this technology by the Forest Survey of India for its regular activities in providing information on forest wealth and regularly monitoring the same. The country is covered by about 190 scenes of the multi-spectral scanner imagery of Landsat, each covering an area of about 185×185 km. It also took some time to scan through all available data, with minimum cloud cover and also of the time period close to the best season for this purpose. So, 1972-75 had to be taken, although theoretically data should be obtained every 18 days, in view of the satellite's periodicity of reappearance at the same place. Similarly, data pertaining to 1980–82 was also used, covering the whole country. Both these sets of data in the form of false colour imagery were visually interpreted to delineate gross, open and dense forest; and maps on 1:1 million scale were prepared covering all the States and Union Territories of India.

According to recent information available from the Forest Survey of India (1993), there has been a marginal increase in the forest cover of 2 200 ha from the assessment made in 1991, using remote sensing data. With the improved type of satellite imagery used now, we can be confident of a real improvement on the ground. It is also pleasing to note that various actions taken on the forest front has yielded some results and have more or less successfully arrested further deterioration. With some more emphasis and effort, it may be possible to slowly improve upon the existing forest cover.

To conclude, the use of satellite remote sensing data has been found to be a reliable and useful tool for gathering forest information. The above study also gives us the approximate rate at which the forest cover is changing and focusses attention on what should be done to ensure sustainability.

10

Microwave Remote Sensing

The Microwave remote sensing system offers certain specific features that complement and supplement the data being generated by the optical sensing techniques.

Earlier, I had made a brief reference to the Electromagnetic Spectrum (EM spectrum) and the range at which the satellite system made observations, i.e. the visible range, the near infrared and the thermal infrared. Mention was made as to how observations made in this range are limited by intervening clouds from the Earth's surface. Such limitations can be overcome by using instruments which can make observations in the microwave regions of the EM spectrum and get data by penetrating through clouds. An editorial in *The Hindu*, dated January 20, 1994 about the earthquake in Los Angeles (US) and a mention of European Remote Sensing Satellite data (ERS-1) (See Plate 10.1) prompted me to write this article on microwave remote sensing systems, which is now becoming an operational reality.

Sensors used in remote sensing satellites are broadly grouped as passive and active, the former also being called optical sensors. The passive remote sensing sensors use the reflected part of the Sun's energy for its observation. The active remote sensing sensors use the energy generated by the sensor itself which may not be visible to us, and obtain data. This overcomes the need for the Sun's energy system. It is also capable of penetrating through clouds and making observations possible at night, except when there is very heavy rain in any region. To some extent, it is broadly analogous to using an ordinary camera with a flash when natural lighting conditions are not adequate to obtain good quality pictures.

The microwave remote sensing system provides certain specific features that complement and supplement the data being generated by the optical remote sensing techniques. The interaction between matter and electromagnetic waves as a phenomenon is now well understood and the use of microwaves at centimetre and millimetre wavelengths as probes, instead of electromagnetic waves in micron wavelengths results in new types of information about the terrain and the surface of the Earth. In the case of microwave remote sensing, it is also possible to use both active and passive sensors.

Instead of using illumination from the Sun, it is possible to use a radio frequency transmitter and receiver system for imaging. Not only can it be used in all kinds of weather and to penetrate through clouds, these techniques can also be used to detect soil moisture, very useful for agriculture, water management, or flood monitoring. In addition, they provide valuable information regarding the ocean in terms of wave heights, surface wind speeds and direction, water vapour content and rainfall over the ocean's surface.

Under the auspices of the National Remote Sensing Agency in Hyderabad, a certain amount of experimental aircraft mounted synthetic aperture radar (SAR) remote sensing and imaging experiments were carried out in 1988–89, using an aeroplane fitted with such a sensor, belonging to a company in Canada and brought specifically for this purpose. Data were obtained covering parts of the Andamans, Kaveri delta, Krishna delta and the Mahanadi delta in Orissa, as a joint exercise with the Oil and Natural Gas Commission, the Orissa State Remote sensing centre, the NRSA, and the Department of Science and Technology. The data were analysed for the various application activities experimentally to gain experience.

In July 1991, the European Remote Sensing (ERS) satellite was launched by the European space agency carrying aboard two specialised radars plus an infrared sensor, a radar altimeter, an along-track scanning radiometer and a wind scatterometer; each giving data for different purposes. This satellite orbits at an altitude of approximately 785 km and in a sun-synchronous orbit, crossing the equator at 10–30 am.

The repeat cycle is three to 35 days, but it can be adjusted to a final 176 days towards the end of the mission. An enormous volume of data is already available and considerable effort is still needed to develop proper application strategies, to make operational utilisation of the microwave remote sensing systems.

A brief description of some of the instruments on board the ERS satellite are given below:

Synthetic Aperture Radar (SAR) image mode: This obtains strips of 30 m ground resolution imagery, 100 km in width to the right of the satellite track. The 10 m long antenna is aligned parallel to the satellite orbit to direct a narrow radar beam sideways and downwards onto the Earth's surface. Imagery is built up from the strength of the return signals which depend primarily on the roughness and dielectric properties of the surface.

SAR Wave Mode: Operations of the synthetic aperture radar in wave mode provides small 5 km × 5 km images (imagettes) at intervals of a 200 km long track. It provides a global sampling of wave spectra suitable for the daily measurement of wavelengths and directions of the main ocean swell wave systems. We in India do not receive these signals at present.

Wind Scatterometer: This equipment uses three sideways looking antennae (beams): one pointing normal to the satellite flight path, one pointing 45 degrees forward and the third pointing 45 degrees backward. Using certain mathematical models one could establish relationship between backscatter, wind speed, wind direction and incidence angle of the observation. Again in India, we do not receive these data at present.

Radar Altimeter: This is a nadir (directly downward) pointing pulse radar used to measure wave height, wind speed and sea surface elevation, the latter of which is appropriate to the study of ocean currents, the tides and the global geoid. We do not receive these data also in India at present.

Along-Track Scanning Radiometer (ATSR): This is a passive instrument with a microwave sounder, consisting of an advanced four channel infrared radiometer providing measurements of sea surface and cloud top temperatures, and a two-channel microwave sounder providing information on the total water content of the atmosphere. We do not receive this data either at present. There are other instruments also, such as the Precise Range and Range-rate Equipment (PRARE), and laser retroreflector.

Ground stations: Countries, i.e. within and outside Europe (including India at NRSA, Hyderabad) will also receive ERS-1 data by arrangement with the European Space Agency (ESA), and this will extend the coverage potential in respect of the high-resolution (30 m) imaging missions with SAR. Four other processing and archiving facilities (PAFs) located in Germany, France, the U K and Italy will be the main centres for the generation of other products and archiving of ERS-1 data and products.

Process of Synthetic Aperture Radar Data: The data/signal received from optical sensors, like the sensors in IRS-1A and B, Landsat and SPOT, can be directly converted into imagery. Corrections and required enhancements of certain features alone need to be carried out to improve the quality of the imagery. However, the data/signal from synthetic aperture radar (SAR), a major microwave sensor on board European remote sensing satellite ERS-1, has to undergo complex signal processing before it can be converted into imagery. The processing time is also comparatively long, although attempts are being made to reduce this and generate imagery in near real-time.

The first European Remote Sensing Satellite (ERS-1) is the forerunner of a new generation of space missions, which promises to make a substantial contribution to the scientific study of our environment. Both global and repetitive observations are needed to resolve the broad range of space and

Fig. 10.1 *ERS-I imagery covering the Vedaranyam area in Tamil Nadu. Some salt pans can also be identified.*

time-scales involved in the monitoring and preservation of the environment (Fig. 10.1). These requirements are very wide-ranging, and Earth observation from space may be the only viable and cost-effective means of acquiring much of the necessary input data for climate models and for monitoring the Earth's surface conditions on local, regional and global scales.

Thus, the ERS-1 is expected to undertake the measurement of many parameters not so far covered by existing operational satellite systems, including those of the sea state, sea surface winds, ocean circulation and sea/ice levels as well as all-weather imaging of ocean, ice and land. The data so far received is now being used to devise operational application activities. Recently, a national symposium on microwave remote sensing and users meet was held at the Space Application Centre (ISRO) Ahmedabad, where a large number of papers came up for discussion on the application aspects.

11

Information and the Space Age

Our entry into space has opened up new vistas in information technology as well. This article offers valuable insights into the expanding potential of this young science.

The explosion in information technology in recent years can, to a great extent, be attributed to the new vistas opened up from space observation platforms and in the rapid advance made in Remote Sensing Technology. There could be several reasons for it, which we will examine here and try and predict all that could happen in the near future in this area. Human beings survive on land and, with the rapid rise in population and the finite amount of land we have, it becomes essential for us to derive the maximum out of the dwindling per capita availability of this valuable resource. All natural resources, renewable or non-renewable, come out of land and its exploration, exploitation and management depend entirely on the information that we have and our capacity to monitor the changes that take place.

Natural resources are inter-related, so any excessive exploitation of a particular resource may lead to deleterious effects on other resources. There is thus a need for an integrated approach when we take up land–water related development schemes. The more real-time the information is, the better will a decision emerge. Since human beings have a limited capacity to process data and information that are available, a means to process them fast is required and this is now available in the form of very fast and powerful computer systems.

When we talk about land, we should relate any information to its true geographical position, or its longitude and latitude values. This has also become possible now with the emergence of powerful remote sensing technology, based on space platforms. This is called the geographical information system, which some people call a system capable of supporting important decisions, by simulating different models in the computer. Thus, we can see that we need to know "what is where and how much of it is available at a particular time". It is space technology in the form of providing data by remote sensing that has added a new dimension "time" in the present day complex information technology.

We also need various means to transfer the information to those who need it in different parts of the world, almost instantaneously. This too has been made possible by communication satellites, placed in orbit at an altitude of 36 000 km. Information on the weather conditions throughout the world is updated by weather satellites. Geographical locations in terms of latitudes and longitudes, are normally provided by land surveyors and geodetic engineers, based mainly on the measurements they make on the ground. During the last decade or so, the US Defence Department have provided a set of 21 satellites which provides latitude and longitude values of a place, using global positioning system receivers, to a precision of less than a metre, based on a new geodetic datum called WGS-1984. Using this technology, it is possible to establish any number of control points on the ground, independently and rapidly, as required for various purposes. This technology can also be used in port management, transport and convoy management, etc.

We are aware of the environmental disturbances that are taking place, the greenhouse effects, sea level rise, etc. all of which need to be studied carefully to identify the causes through understanding the global changes and global processes that are taking place. For all these activities, substantial input data is derived from space platforms and various remote sensing data. The international global change monitoring system is itself designed on information derived from space platform data systems. Space technology has thus opened up new vistas in accelerating the development of information technology in its totality.

In this context, it may be appropriate to briefly describe the technology of satellite remote sensing and its various applications and also the geographical information systems, that have become a popular tool in decision-making.

Geostationary satellites are used mostly for communication and to some extent to monitor weather. There are satellites specifically designed for obtaining various meteorological parameters. Satellites that are used for remote sensing orbit the Earth in a predetermined manner. Most remote sensing satellites orbit at an altitude ranging between 700 and 1 000 km in a north-south direction, somewhat inclined to the polar axis. They carry sensors or scanners or linear imaging scanners with charge coupled devices or in some cases high-resolution long focal-length camera systems. They make use of the reflected part of the Sun's energy and record the reflectance of various objects on the surface of the Earth in digital format in terms of grey levels.

These digital values are beamed to the Earth, where a tracking antenna receives the data and records the information on high density digital tapes.

From these tapes, it is converted to normal computer-compatible tapes for further processing, before they are made available to users. The digital data may be directly used in an interactive computer system for various kinds of analysis and classification. This digital data can also be converted to analogue data in picture format, using optomechanical or optoelectronic photowrite systems, which can then be used for visual analysis.

The scanner in the satellite can be designed to record data from the Earth in several channels, converting different parts of the electromagnetic spectrum, which can be combined as desired. Since the situation as it exists on the surface of the Earth when the satellite passes over it is recorded, we are able to get a replica of the ground that can be studied in the office either using computer-aided systems or simple visual-aided systems, and interpreting various details as we desire and delineate them. A good geologist can interpret and prepare a geological map, a soil scientist will be able to delineate various soil classes and a land use expert can classify and demarcate various land use classes. All using the same data. Before finalising the classification, a brief ground truth verification is carried out and appropriate changes are made. This process is very fast compared to what we normally do by conventional ground methods.

The scene of such a satellite data may cover anything between 3 600 sq km and 30 000 sq km, depending upon the design of the sensor and the spatial resolution it provides. Since the Earth is also rotating along its polar axis, these orbiting satellites reappear covering the same area once in say 16 to 26 days, depending upon the design, thereby giving us data and information on a regular basis. Since each scene covers a large area, we also get a synoptic view, which otherwise is not possible. Thus, we are able to obtain near real-time information of various happenings on the surface of the Earth, periodically. However, till recently, one of the shortcomings of this satellite data was that the data could not be got on cloudy days. But in the last two years, this has been overcome by microwave remote sensing data, using active sensors which generate their own energy systems. The first such satellite is in orbit, launched by the European Space Agency, and we have the facility in India or rather receive data when it passes over our country. The Shadnagar station near Hyderabad can receive data from such remote sensing satellites from all areas within a 2 500 km radius data covering some of our neighbouring countries.

The evolution of the National Natural Resources Management system under the Department of Space, established nearly a decade ago, has come a long way in providing information to planners and other users. There are five regional remote sensing service centres working with enormous digital analysis capacity. There are also some state remote sensing centres, like in Uttar Pradesh and Tamil Nadu, where similar large and powerful computer systems are working to create information about our land and its resources. In addition, almost all states have their own remote sensing application centres, and all these contribute to the National Natural Resources Management System. The Natural Resources Information System has also taken root in the Department of Space and is rendering valuable service.

A number of private industries and entrepreneurs in the market support all these activities. Many private firms are even considering launching satellites with sensors on board to get the information they need. However, the launch of such satellites is still controlled by the Government in most countries. The CIA Senate Committee on Intelligence and the US State Department are seriously considering the open sale of spy satellite products, with a spatial resolution of one metre. The Russians are already selling their reconnaissance spy satellite data of two metre resolution in the open market. If privatised, there is scope for a market of $5–15 billions in the coming years. Some of the other commercial values of satellite remote sensing information have been profitably used in the area of agriculture and ensuring food security at economic price levels. In our country, we have been able to estimate the area under major crops every year using our own remote sensing satellite data and give an estimate of the total yield within certain levels of reliability. This information can be made available just before harvest, which can be used by our planners to devoid upon the procurement strategy or excess, i.e. whether to import or export and if so at what price and from where, and to maintain a secure enough level of food to feed our large population.

Another innovative application in the US has been by the chain of fast food shops. They have been regularly using satellite remote sensed data to decide upon the locations of new shops in various parts of the US. Another application could be in the area of real estate business, which is growing very fast throughout the world. Satellite data can be used to take decisions on potential development areas, keeping in view the trends. Some time ago, even the media agencies in the US and Canada were contemplating designing and launching a satellite called "Media Sat" which was supposed to provide high resolution information to the news agencies, which could attract the attention of the public. Some of the big oil companies in the world are also thinking of launching their own satellites, but presumably it has not yet taken shape.

A geographic information system (GIS) allows us to create, maintain and query electronic databases of information normally displayed on maps. These databases are spatially oriented, the fundamental integrating element being their position on the Earth's surface. This system consists of a set of computerised tools and procedures that can be used to effectively encode, store, retrieve, overlay, correlate, manipulate, analyse, query, display (both graphically and numerically) and disseminate land-related information. They also facilitate the selection and transfer of data to application specific analytical models capable of assessing the impact of alternatives on the chosen environment. The underlying foundation of a sound GIS is an effective digital map database, tied to an accurate horizontal control survey framework. As we move into the information age, the value of data and information as a marketable product is directly related to the quality of data. This quality is based upon such things as reliability, consistency, accuracy, real-time, and how easy it is for others to access, integrate and use the data.

In India, the long term programme of the Department of Space gives an assurance of continuous availability of satellite data, without interruption, in the future.

There are several companies in the world now adding value to the raw remote sensing data and are growing fast in commercialising creation of information products to provide solutions for their clients. In India, this activity is yet to develop in a big way, although the Department of Space of the Government of India is encouraging its own employees to take up such challenges by providing assistance, in various forms. As in all cases, there is enough scope for private entrepreneurs and industry to enter this field in India, and create user-oriented information systems and solutions, using space technology, which is freely available without any restrictions. The future remote sensing satellites are expected to provide much better resolution data as well as being oriented towards specific needs.

12

Remote Sensing for Route Alignment

Remote sensing aids us in planning pipeline routes, ring roads and urban settlements.

Remote sensing as a tool can be used in certain specific cases, with considerable advantages both in terms of cost and time, with acceptable standards of reliability. In short, whatever information is required to be generated about the ground conditions on the Earth's surface, which we normally do by long visits to the ground and by several people, is now possible by using remotely sensed data, as the terrain condition as it appeared at the time of satellite sensor observation can be studied in the office with minimum visits to the ground for only verification purposes.

Investigation activities in many projects such as road alignment, railway alignment, canal alignment or inter-basin river water transfer and pipeline alignment takes considerable time if we have to depend on traditional methods, by studying the topographical maps and several visits to the site.

Topographical maps are general purpose maps, and several details on the Earth's surface are symbolically depicted. We may not get an objective view of the terrain in its true form, particularly when these maps are too old. Generally speaking, most of our Survey of India topographical maps could be as old as 25 to 30 years, and so, do not depict the ground conditions as at the time when a study for a new road or pipeline alignment is undertaken.

For such studies, we also have to determine the obstructions that have to be avoided, prevent cutting of large forest areas, and ensure the least environmental impacts. For example, if we have to carry out a study as to how best we could link various peninsular river systems, we could do the pre-feasibility study in a much shorter time and at less cost using the latest available satellite remote sensed data, than by any other methods, as satellite data or imagery covers a large area in one scene, which enables a synoptic view of the entire area. Today, there are no better data systems that can match remote sensing data. Even for feasibility studies, the use of such data will be advantageous, for it will eliminate a number of uncertainties.

In the case of route alignment studies there will have to be a good analysis done of the morphological information covering the area, besides the

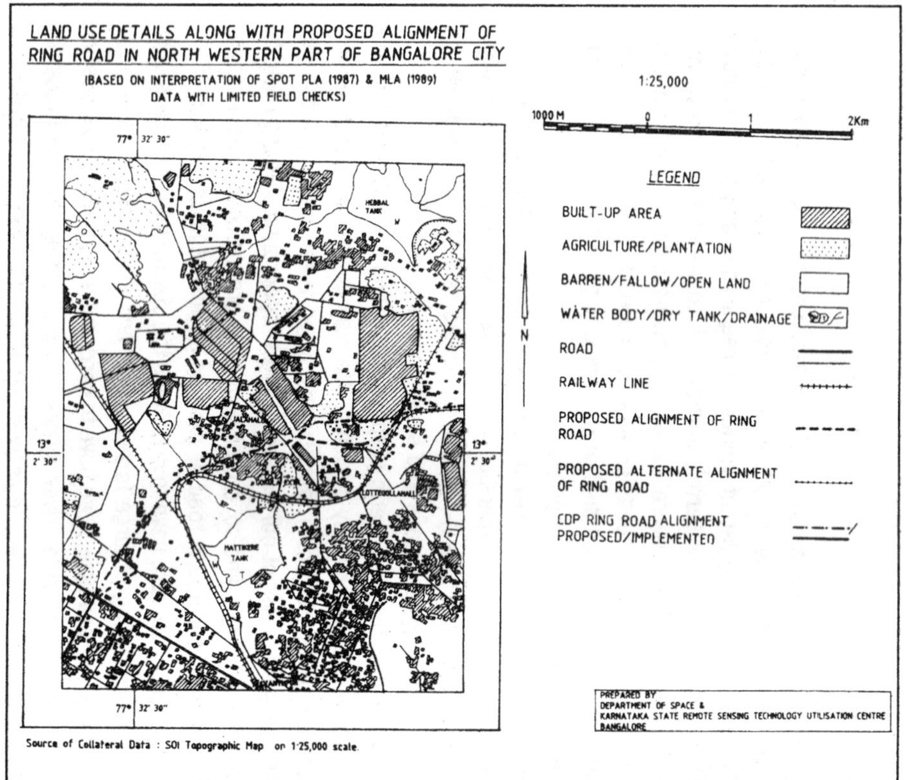

Fig. 12.1 *Map prepared from a satellite image for pipeline alignment (Courtesy NNRMS-ISRO, Bangalore)*

general geological and terrain aspects. In addition, the approximately selected corridor will have to be free from causing undue environmental impacts, and free from too many obstacles. Such varied and complex information needed for a final selection may take considerable time if we resort to conventional ground survey methods. To obtain an initial overview of large areas to determine the corridor likely to be suitable for further detailed study, satellite based remotely sensed data are the best. However, this effort will in most cases be supported by using aerial photography to narrow down the most likely routes before design activity starts.

To illustrate the above concept, an oil pipeline alignment study carried out by the Geosciences Division of the National Remote Sensing Agency (NRSA) to connect Haldia to Budgebudge in West Bengal as required by the Indian Oil Corporation Limited (IOCL) in 1992 is described here. The data used were from an Indian Remote Sensing Satellite Liss II Camera, with a ground resolution of approximately 36.50 m and the French SPOT satellite data with a better ground resolution of about 10 m. The interpretation of this data was done to cover habitation, forest cover, slope, trends of flooding in major and minor rivers in the area, types of soil, etc. to arrive at the shortest alignment.

These studies resulted in producing

a) a map showing pipeline alignment and the inhabited areas and

b) a map showing pipeline alignment with drainage, canals, roads and railway-line crossings (Fig. 12.1). These maps were on a scale of 1:50 000. The total distance covered was of the order of 60 km. The maps were checked on the ground by visits, to ensure that no errors had crept in during the interpretation of the remotely sensed data.

The IOCL has taken full advantage of the information made available by NRSA to select the most suitable alignment, considering the various engineering and other constraints. It must be emphasised here that the study took only about four months, otherwise it could have taken at least one year.

The interpretation of satellite imagery was done by using simple visual aids, by some experienced geologists of NRSA. The various aspects they considered were:

Occurrence of pucca roads, railway and inhabited areas to consider approachability with minimum crossings.

Potential hazards due to nearby river systems being heavily flooded during the rainy season.

Number of river crossings that may be required.

Good forest cover, sanctuaries, reservoirs, etc. that may get affected.

The type of soil cover for establishing stability.

General level of the ground along suggested routes.

Possible environmental hazards to population in case of any mishap due to possible bursting of the pipeline.

Armed with all this information above certain alignments were suggested to the IOCL, who could then decide on the final pipeline route.

In another case of selecting a proper alignment for a ring road around Bangalore city, particularly along three difficult stretches as needed by the Bangalore Development Authorities, the Karnataka State Remote Sensing Technology utilisation along with the scientists of the Department of Space, Bangalore carried out a study, using high resolution remote sensing data obtained from the French satellite SPOT and provided a viable solution in about a month. Both visual and computer-aided interpretation of remotely sensed data were used for this study, supported by available aerial photographs on the 1:10 000 scale of 1987, and the topographical maps on the 1:50 000 scale. Brief ground visits were also made for verification.

The criteria adopted included items like keeping the road short, easy and safe with regard to manoeuvrability, avoiding obligatory land use features like

dense settlements, religious places, following the topography to balance earthwork, avoiding areas prone to erosion and loose soil, to keep the number of blind curves, and drainage cuttings to the minimum. Several maps were prepared and submitted along with a short report to the Bangalore Development Authority to enable planning the ring road most optimally (Plate 12.1).

These two examples illustrate the utility of remote sensing in route alignment studies, to economically and speedily arrive at the best possible solutions. In the coming years, remote sensing technology is poised for providing data with better ground resolution and capable of meeting more stringent information needs.

13

Remote Sensing for Flood Management

Remote sensing techniques have been found to be highly promising in flood damage assessment, for it can be carried out within a week of the occurrence. The country should plan and prepare itself to get data for areas that are likely to be affected.

The occurrence of floods in India is an annual feature. While we should be prepared to manage a situation where water becomes scarce, we should be equally prepared to meet a situation caused by excessive water in the form of floods. Floods are mainly caused by heavy rain associated with deep depressions (low pressure) and cyclones in areas having inadequate drainage. In India, it is estimated that floods annually affect seven percent of the land area, the most affected areas being Assam, Bihar, Uttar Pradesh and West Bengal and at times Andhra Pradesh, Tamil Nadu and even Jammu and Kashmir.

Floods cannot be totally controlled, but the extent of damage caused can be reduced to a significant extent by taking timely and appropriate management measures, both structural and non-structural. The structural measures include channel improvement, construction of embankments, dikes, bunds, diversions and bypasses, reservoirs and retarding basins. The non-structural measures are based on flood-risk zoning, flood forecasting, relief and timely rehabilitation of people affected. The total loss on account of flood damage to crops, houses, cattle and public utility has been of the order of Rs 26 800 crores, during the period 1953–1987.

The chronically flood-prone river basins are Ganga and Brahmaputra. These Himalayan rivers flowing down the mountains cause severe flood problems almost every year in Uttar Pradesh, Bihar, West Bengal and Assam due to high discharges concentrated during the monsoon months (June to September) and large volumes of sediment which are carried to the plains. Most of the flood plains of these river basins are densely populated with the result that these people are the worst victims almost every year.

For planning flood control measures, timely, reliable and latest information is needed and it is here that remote sensing plays an important role, and there are no other means known to us at present. Conventional forecasting techniques based on antecedent precipitation over the catchment give forecasts 24 to 48 hours before a flood. Techniques have also been developed to monitor bad weather conditions and rainfall estimation using satellite data, so as to forecast flood to some extent in advance and take precautionary measures to minimise losses. Remote sensing techniques have been found to be highly promising in flood damage assessment for they can be effectively carried out within a week's time of the event.

Timely and immediate relief to the affected gets priority. To optimise relief operations and its systematic management, a flood inundation map is a vital input. For this, data obtained from satellite remote sensing has been found to be the quickest and most powerful medium of information. With data now available from several remote sensing satellites such as IRS-1A IRS-1B, American Landsat, French SPOT and European Remote Sensing Satellite (ERS) it has become possible to monitor flood situations most effectively so as to enable speedy decisions. It can provide the approximate duration of flooding if information on the daily water level discharge is available. Decisions made using flood inundation maps have been found to be efficient and highly cost-effective and also give a better idea of the damage in terms of its cost.

The National Remote Sensing Agency (NRSA) has mapped major flood situations during the last decade or so, using near real-time satellite remote sensed data and gives that information to the departments and officials concerned. Moreover, certain state remote sensing applications centres, like UP and Andhra Pradesh, have carried out flood mapping during the past several years and are now well geared to tackle these situations in the shortest time possible. Such mapping activities from satellite remote sensing data, have been seen by top officials and political leaders to make rapid assessment of relief assistance needed, except in very specific cases where continuous cloud cover over the affected area persists for an unusually long time. Even in such cases, now with the availability of European Remote Sensing Satellite (ERS) data, the situation is improving gradually.

The NRSA keeps a close watch on the flood situation in the country through the news media, television and radio reports, and satellite data is screened for cloud cover covering the flood affected area. Even when partially cloud covered data becomes available, it is subjected to rapid visual interpretation supported by digital enhancement and analysis. In addition to the monitoring of the floods regularly, based on the acquired satellite data during floods, pre-flood and post-flood situations and with the available ground information, an approximate assessment of flood damage is made by the scientists of NRSA, which is then profitably utilised by different Government departments.

In September 1992, severe floods occurred in Jammu and Kashmir, causing extensive damage. Scientists of NRSA, Hyderabad, immediately took note and obtained cloud-free satellite data of September 13 covering the Srinagar Valley (Dum river from Ananthnag to Baramulla) soon after the floods (Plate 13.2 & 13.3). They also obtained pre-flood satellite remote sensed data of 1991, the earliest available (Plate 13.1). These data were digitally and optically enhanced and enlarged to a scale of 1:50 000.

The interpreted figures and flood maps were prepared almost on real-time basis. It was a coincidence that on September 13, 1992 ERS (Microwave) data was available covering the same area, enabling a better information development. Further to this, data obtained again on September 24 was also studied for the post-flood situation.

Table 13.1 - Flood Statistics: Srinagar Valley, Jammu and Kashmir

Area under standing water as on September 13, 1992	178 sq km
Crop area under water as on September 13, 1992	150 sq km
Marooned villages	220
Length of bunds in standing water area	72 km
Standing water as on September 24, 1992 (post- flood)	55 sq km

The unprecedented cyclone of May 1990, which hit the coastal districts of Andhra Pradesh, affected vast areas of standing crop and public utilities. The IRS data of May 11 recorded the cyclone and other flood features, despite the constraint of cloud cover. These images clearly show the areas which were affected, the crops that were wiped out, the roads, the canals, the railway tracks, etc. All this information provided valuable input for estimating total damage. But for such satellite remote sensed data, it would not have been possible to make any reasonably reliable estimate of damage speedily.

Similar studies have been undertaken in other areas prone to floods on a regular basis, such as the Kosi, Ganga and Brahmaputra, almost every year when floods occur. This technique has been almost operationalised and NRSA has been making attempts to transfer this technology to various state remote sensing applications centres as well as to other government and non-government agencies concerned with managing floods. Future use of satellite remote sensing data in such flood studies are highly promising, particularly with improved spatial resolution data as is being planned in the IRS-1C to be launched in 1995.

We should plan and prepare ourselves to obtain data of possible areas likely to be affected as, in our country, we have noticed severe flooding even in new areas, a thing not anticipated in the past. To sum up, it can be stated that remote sensing is the only technology available, as of today, for it provides timely, accurate, reliable and periodical information on flood inundation and damage, in the shortest possible time.

14

Remote Sensing and Geographical Information System

The Geographic Information System (GIS) help us manipulate data in the computer to simulate alternatives and to take the most effective decisions.

We have seen how satellite based remote sensing provides data on various natural resources of our planet Earth and how this data could be converted to obtain information on such resources, giving us an idea of its extent and condition. Besides, this information being available to us at periodic intervals enables us to monitor changes that occur in a dynamic environment and focus attention on remedial measures required to arrest deterioration of the environmental condition or at least minimise it.

While we can see large areas on the surface of the earth covered by one scene, it was also brought out that data is available in digital format which could be analysed for extracting information by using computers. In addition, different types of information on natural resources such as water resources, groundwater potential, soil survey information, land use, vegetation information, etc. can be extracted from the same satellite data. (Fig. 14.1)

With the availability of digital information related to geographic positions, with precision, it can be used to orient various natural resources information to their actual location on the ground, thus enabling us to know "what is where" and how much of "what is where". Since all natural resources are inter-related in some form or the other and since we have to take into account other forms of secondary information before detailed exploration and exploitation activities, we need to have an integrated picture, to enable understanding what effect certain resource potential will have, when we decide to exploit some other resource.

For example, if we over exploit groundwater near coastal regions, the aquifers will get brackish due to seawater intrusion. Similarly, if we overexploit forests, it will result in a certain impact on the ecosystem and the environment itself. It is therefore necessary for us to understand what side-effects may result in our decision to launch any development scheme, may it

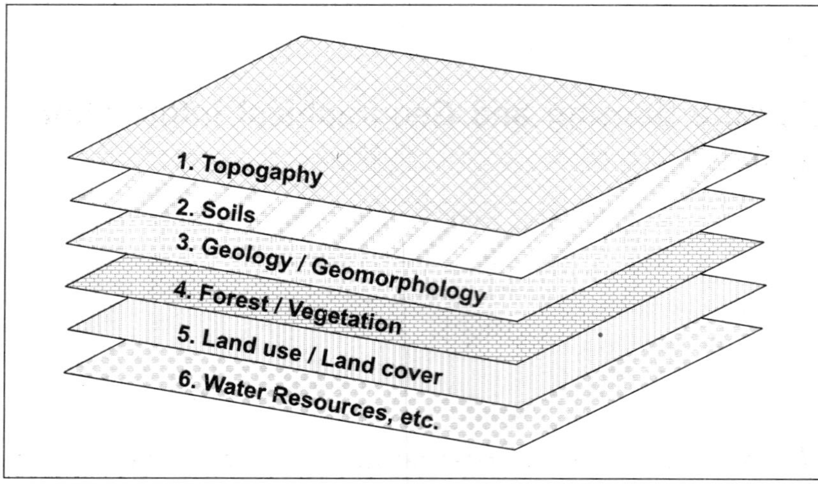

Fig. 14.1 *GIS data organisation.* (*Courtesy: ISRO*)

be irrigation or aquaculture or dam construction for power generation, and so on.

Since all factors affecting the exploitation of various natural resources can be made available in the form of digital values, and spatial locations well defined with reference to the true geographical position on the surface of the Earth, we are now in a position to manipulate these data in the computer to simulate various alternatives and give us an idea of the consequences of our decisions and thus enable taking the most appropriate or effective decisions to obtain optimum results.

This type of activity through use of computers, where all types of data are stored in different layers in a graphic form and used for different combinations is generally termed as a geographic information system (GIS). In short, it is an information system used for management decisions and allows us to create, maintain and query electronic databases of information normally displayed on maps. These databases are spatially-oriented, the fundamental integrating element being the position on the Earth's surface.

GIS consists of a set of computerised tools and procedures that can be used to effectively encode, store, retrieve, overlay, correlate, manipulate, analyse, query, display (both graphically and numerically) and disseminate land-related information. "Further they facilitate the selection and transfer of data to application specific analytical models capable of assessing the impact of alternatives on the chosen environment."

The main difference or factor which separates GIS from other information storage and retrieval systems is the spatial and specific locational features given in a co-ordinate form, for reference and use as an important variable in quantitative analysis. "The spatial data is generally in the form of maps, which could be showing topography, geology, soil types, forest and vegetation, land use/land cover, water resources availability (both surface and ground water) etc., and stored as layers in a digital form in the computer."

Besides spatial data, we could also have attribute data, like statistics, texts and tables. New maps can be easily generated by integrating many layers of data in a computer. Thus, a GIS has a database of multiple information layers that can be manipulated or analysed to evaluate relationships among the selected elements, from all that is available.

When we look at the process of making a modern topographical map, it starts from a series of aerial photographs of the required landmass in a predetermined manner and from specific altitudes to meet the standards of scales required of the ultimate map. We then join these photographs together and trace the features we want to include on our map directly from the photographs.

This appears simple, but actually it is not so, as the aircraft tips up and down, is blown sideways, and tilts from side to side from time to time, so the photographs that were taken have to be corrected or adjusted to eliminate all the distortions caused. For this we need accurate ground control points in terms of latitude and longitude values, fixed on the ground and which could be identified in the photos.

This procedure of ground control surveys gives us the x and y coordinates for each of the points. In recent years, this procedure of obtaining ground control points has become possible using what is called Global Positioning System (GPS), which dramatically reduces time and cost. The GPS is based on a set of 21 satellites, put in orbit in a predesigned manner, signals from which give us the co-ordinates of the point where such receivers are kept at a very high accuracy. No detailed description of the use of GPS is attempted here.

With the above mentioned framework of precise geographical coordinates and details, we can superimpose on this cartographic framework digitised geological map information, hazard areas, flood plains, soil types, contours and so on. Once things are co-ordinated and referenced to the same framework and in digital form, it is amazing what one can do with them, if we have the right kind of GIS software. Any feature on the surface of the earth can be depicted on a map by either a point, a line or a polygon (closed in shape). However in the case of a real GIS, other types of attribute data or alphanumeric or tabular information can also be integrated.

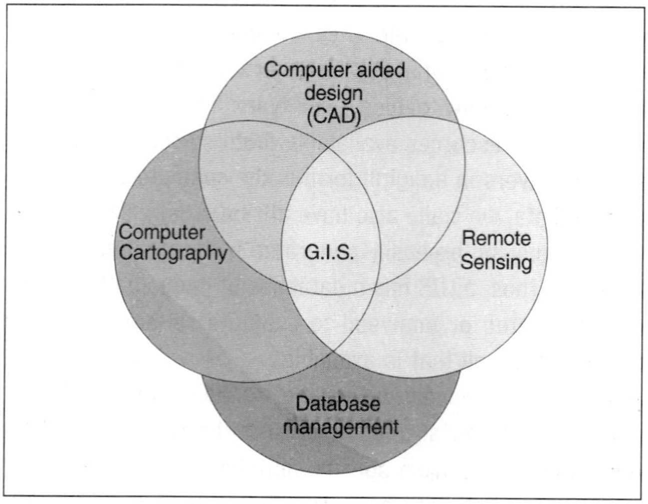

Fig. 14.2 *Characteristics of several related systems.*
(*Source: Concepts from John Kerr and Gabrieles Zillmer.*)

Thus, the versatility and capability of a GIS software system very much depends upon how attribute information is related to the spatial entities they are associated with, as well as the way in which the spatial entities are related to each other.

Yet another aspect that we should recognise in a GIS is the geometric relationship or spatial entities, i.e. points, lines and polygons that are dependent on adjacency rather than upon their locations in space. This aspect is termed as Topology. In fact, the capability to handle topology is what differentiates a GIS from a graphic system.

Topology refers to the ability of the system to create, maintain and work with relationships between features. We should therefore recognise the differences between ordinary simple graphics or CAD/Cartographic systems and a true geographic system, which is a complex data structure utilising topology. In the case of a usual cartographic graphic system, the emphasis is on display and the main objective is to produce maps. Whereas in GIS, emphasis is on retrieval, analysis and decision support, and it can generate new information. To summarise, GIS is made up of all characteristics of several related systems, viz. digital cartography, computer aided design (CAD) remote sensing and database management, to obtain new information (Fig. 14.2).

It has also been recognised all over the world that a good background in land

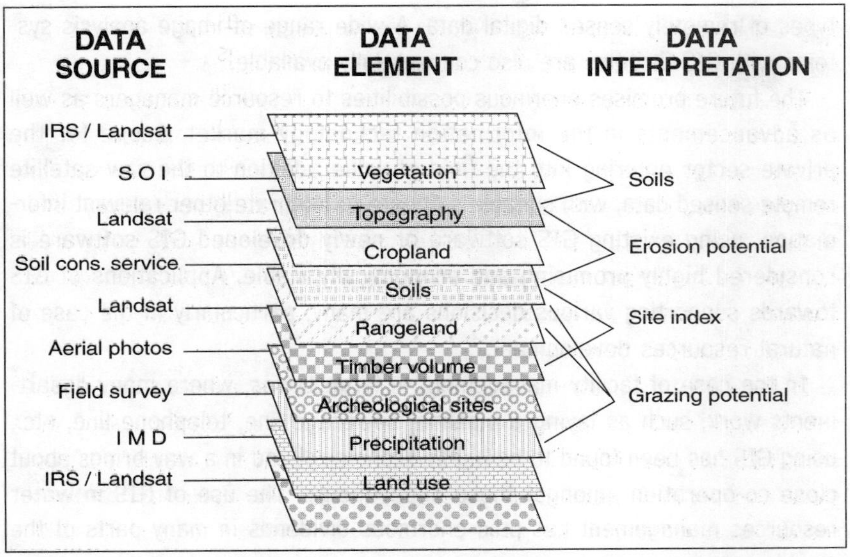

Fig. 14.3 *Integration of Landsat data with other sources to derive data elements input to a GIS interpretation model. (Courtesy: ISRO)*

surveying/survey sciences and mapping can be of great advantage for implementing a GIS programme, because Geographic Information Systems are generally structured on horizontal control survey networks, digital topographical databases, and many times on surveyed land parcel databases (Fig. 14.3).

In all cases of establishing a GIS, it is essential to have a user needs study so that the entire operation could be user driven. In fact, a proper user study itself will have to be carefully planned and may take some time. Besides this, to maintain consistency, we have to evolve certain standards agreed upon by all types of users of the facility being set up. The quality of data should also be reliable, consistent, accurate and easy for other to access, integrate and utilise.

Remote Sensing in GIS

As already mentioned, remotely sensed data is one of the main sources which provides information on many land and water related subjects, on a near real-time basis. Since such remote sensing data is acquired in digital format, it is possible to carry out computer aided interpretation of such data using various kinds of software already available.

Such information is therefore easy to feed into a GIS environment to be integrated with other types of data. It can be stated that digital image processing techniques enable speedy and accurate interpretation of the various types of remotely sensed digital data. A wide range of image analysis systems with GIS facilities are also commercially available.

The future promises enormous possibilities to resource managers as well as advancements in the information technology market. Scope for the private sector entering into the field of value addition to the raw satellite remote sensed data, with suitable software to integrate other relevant information, using existing GIS software or newly developed GIS software is considered highly promising and economically viable. Applications of GIS towards supporting various decisions are many, particularly in the case of natural resources development.

In the case of facility management in large towns, where many departments work, such as laying a pipeline, sewerage line, telephone line, etc., using GIS has been found to be highly economical and in a way brings about close co-operation amongst these departments. The use of GIS in water resources management has paid enormous dividends in many parts of the world.

The use of GIS as a decision support system is becoming more and more popular. Even certain consumer articles manufacturing industries may find it useful in their marketing effort to reach the maximum number of people economically. It is believed that Madras has a good cartographically accurate base map on a large scale of 1:3 000 available. If the authorities decide positively, a good GIS can easily be achieved economically. One must simply have the will to do it.

15

Monitoring Urban Growth Using Remote Sensing

Cities are fast becoming overcrowded, necessitating a planned and regulated urban growth. For this we require a real-time data gathering system to map and monitor the urban sprawl. Remote sensing offers excellent possibilities in solving this problem.

We have been witnessing a gradual and steady transformation from a predominantly rural society to an urban society in India. The level of economic progress is seen in the growth of urban areas and cities, which serve as nerve-centres of developmental activities. This percolates down to the rural areas. Cities are getting overcrowded and expanding in an unorganised manner and are difficult to control due to regular rural migration to the cities, apparently in quest of better opportunities.

It is understood that the urban population has increased from 25.85 million in 1901 to 217.18 million in 1991. It has almost doubled between 1971 and 1991. By the turn of the century, over 30 percent of India's population (about 310 million) is expected to live in cities or towns. There will be 40 cities with a million plus population as compared to the current 23 or so. All this growth has only increased the strain on available services and led to a gradual degradation of the environment. This has necessitated a planned and regulated urban growth, for which we need a near real-time data gathering system to monitor change using modern technological tools. Remote sensing offers excellent possibilities to map and monitor urban sprawl to a great extent.

While the all-India average growth rate in the decade 1981–91 has been 36.19 percent, the corresponding growth rate in the Class I cities have been 46.87 percent. In fact, some of these cities such as Visakhapatnam and Hyderabad have registered decadal growth rates of 74.27 percent and 57.48 percent respectively (R P Agarwal). Planning for such growth rate essentially needs the most up-to-date and accurate physical maps. Planning exercises will become purposeful when the decisions are based on sound land, social and environmental information. It is in this area of providing physical inputs that remote sensing (both aerial and satellite based) has shown the way. It is therefore necessary that planners and administrators realise the importance of various types of remote sensing data available for urban planning.

For regional planning, data on (i) physical features, i.e. topographical features, geological base and minerals, soil associations, drainage pattern, vegetation, water bodies and climatic conditions, (ii) economic base, i.e. agriculture, forestry, industry, construction materials, commercial and (iii) transport and communications network, condition and quality of roads and expected traffic volume, are all required. In the case of town and city planning, data relating to built-up areas is important. Major categories are: (i) land use in respect of residential, industrial, commercial, recreational (ii) housing with respect to quality, density, types and structure, dwelling unit size etc. and (iii) demography in terms of population density, occupational patterns, etc.

All the above mentioned items suggest that remotely sensed data with its high spatial resolution will be desirable for urban studies. The spatial resolution in this case may depend upon: a) size, b) density and c) contrast of features present in the urban environment. Besides spatial resolution, one should think of an appropriate temporal resolution to monitor rapid changes that occur in an urban environment.

Table 15.1: *Data availability*

Sr.No.	*Date*	*Scale*	*Resolution*	*Remarks*
1.	Landsat TM	1:1000 000	30 m	Can be enlarged to 1:50 000 scale
2.	SPOT panchromatic	1:400 000	10 m	Can be enlarged to 1:50 000 scale and (sometimes even upto 1:25 000 scale)
3.	SPOT multi-spectral	1:400 000	20 m	Can be enlarged to 1:50 000 scale
4.	IRS-LISS II	1:500 000	36.5 m	Can be enlarged to 1:50 000 scale

Table 15.1 gives us an idea of the type of satellite data that is available in India, which has been used for studying the urban growth of some of the important towns an cities during the past decade or so, which includes the metropolitan cities also.

Besides the above satellite-borne remotely sensed data, we could use aerial photographs of any desired scale within practical flying limitations. Aerial photography is expensive compared to satellite remote sensing data, and particularly so when we have to regularly monitor changes that occur.

When we consider the availability of much higher spatial resolution satellite data, during the past 4 or 5 years, the Russian space data at a spatial resolution of upto 2 m may be procurable, covering an area of interest as these are now and being marketed by certain agencies at affordable prices. The advantage in this case would be that, as of now, such data usage is not covered under the security classification of the Government of India, as in the case of free usage of aerial photography, although such high resolution data cannot match the versatility of aerial photographs in all its aspects of mappability, to produce very precise and positionally accurate maps. However, they could form an excellent tool for monitoring changes in urban situations, if they could be procured at regular intervals of time.

The situation in the availability of higher resolution data (better than 10 m) is almost now assured in India, when the latest version of the Indian Remote Sensing Satellite 1-C is launched in another year's time, for such urban mapping activities. While topographical maps on the 1:50 000 and 1:25 000 scale are generally available to be used as base maps and as collateral data, it would be necessary and highly desirable to have base maps of scales of 1:10 000 or even larger for urban planning activities, which at present does not seem to be available, except perhaps in very few cases, although efforts are being made to produce them to cover a larger number of cases.

Aerial photography to cover about 55 towns in Andhra Pradesh, about 35 to 40 towns in Tamil Nadu and also to cover some important cities like Bangalore, Madras, parts of Bombay, Delhi, etc. by NRSA on scales appropriate for preparing large scale maps in those areas, a few years ago. In addition, a new concept of using small aeroplanes fitted with high resolution 35 mm or 70 mm format cameras for monitoring changes is becoming popular, as this technique is about 5 to 10 times cheaper than normal aerial photography, depending upon the circumstances. Processing of such small format aerial photographs (SFAP) can also be carried out using less expensive instruments.

In addition to what has been mentioned above, there is already a move in countries like the US to offer even 1 m resolution reconnaissance satellite pictures for open sale and to privatise producing such data by large companies who may come forward to launch such satellite systems.

While we have seen the above possibilities, let us examine a few examples as to how we have carried out urban area studies and produced information and maps to enable urban planners to use them, based on the type of satellite remote sensed data as tabulated in Table 15.1. In the year 1990 or so, the scientists of the Space Application Centre (SAC)-ISRO carried out an urban land use mapping of Ahmedabad city using data obtained from different satellites, such as the American Landsat, French SPOT and Indian Remote Sensing satellites. The work was supported by topographical maps available on 1:50 000 scale. The work was carried out using data available in March 1975, January 1982, December 1986 and January 1989, to determine the growth of this city over a period of about 1-1/2 decades. The city's growth was found to be from 124 sq km in 1975 to 285 sq km in 1989, which is more than double. It was also found that the higher the Spatial Resolution (SPOT), the clearer the discernibility of various features normally found in large cities and towns.

In another case, scientists of SAC made an urban sprawl map of Bombay city and also a land use map of Greater Bombay and the New Bombay Region, showing the spatial distribution of various land uses (See Plate 15.1). In this case, both visual and digital techniques were adopted in using satellite remote sensed data of different types, based on availability and supplemented by topographical maps. It was noticed from the urban sprawl map that the area had increased from 310 sq km in 1983 to 505 sq km in 1987. A further study was reported in 1991, on similar work done by scientists of SAC and that of the Calcutta Metropolitan Development Authority, using satellite data. In this case also it was confirmed that remote sensed data gave reliable and near real-time information on various aspects concerning development of a large city like Calcutta and such information could easily be used to carry out trend analysis for future growth and plan for growth in a more orderly manner.

Similar work was carried out by NRSA, Hyderabad covering large cities like Madras (See Plate 15.2) and Hyderabad and the results were highly promising as an important tool to be used by city planners very effectively. Based on such techniques, a master plan for proposed land use of non-municipal areas of Hyderabad was prepared and was discussed at length in a recent seminar organised by the Hyderabad Metropolitan Development Authority.

With the increase in population, cities and towns have started growing rapidly. Conflicting interests leads to taking important decisions on the use of the land in a particular area.

It is essential to know the consequences of such decisions and if possible find alternative courses open.

For all such activities, we need near real-time information to create a geographical information system, so that decisions taken will be more appropriate and ensure the orderly growth of a city. The problems faced by city and town planners are becoming more and more complex and resorting to advanced technologies is not only desirable but has become a necessity. It is also envisaged that by the turn of the century, more remote sensed data is likely to be available with much better spatial resolution, ideally suiting the needs of city and town planners.

It should be mentioned here that today's problems and situations are different to those of yesterday's and we should resort to today's technologies to tackle them, rather than follow older methods and procedures, leading to better solutions.

16
Specialised Application of Remote Sensing

A specialised application of remote sensing is in the field of Defense, both for gathering intelligence information and for spying.

Obtaining information or intelligence is one of the major activities of defence forces in all countries to formulate strategies and for long- and for short term planning. During peacetime, the defence forces prepare themselves for war and during the war they fight to bring peace.

Defence preparedness is a continuous process and it cannot be left to chance. In every country, the Army, the Navy and the Air Force, all have their own departments which are actively engaged in matters connected with defence intelligence. The methods adopted may vary from time to time and from country to country. Activities such as spying and counterspying are common. Continuous evaluation of our strategies and plans are carried out by professionals, to ensure maximum security to the country from possible aggression or hostilities from any quarter.

For gathering intelligence, we could resort to ground methods, aerial reconnaissance and data collected from space platforms and satellites. The use of ground methods and aerial methods for such purposes covering transfrontier areas has certain amount of risks that go with it. Use of space platforms and satellite remote sensing appears to be less risky as of now, but needs a very high level of technologies and resources. In the early 1960s, only the two superpowers (US and USSR) were known to have been engaged in very high level space technologies.

We are aware of the detection of missile sites in Cuba by an American U-2 camera flown at high altitude and which produced the pictures as evidence to the USSR. In the case of satellite remote sensing, the data could be considered as useful for natural resources evaluation and management; this is one side of the coin, and making use of it for collecting strategic information is the other side of the coin. In this context we shall examine how satellite remote sensing or in general how space activities can be used by military authorities.

The term remote sensing is most commonly used in connection with electromagnetic techniques of acquisition of information. These techniques utilise a wide range of electromagnetic spectrum (EM-spectrum) including visible, infrared, thermal, and microwave regions of the EM spectrum. The passive or optical remote sensing uses the reflected part of the Sun's energy, while active remote sensing uses the energy system developed by itself.

Target	*Detection*	*General Identification*	*Precise Identification*
Bridge	6	4.6	1.5
Supply-Dump	1.5	0.6	0.3
Airfield facilities	6	4.6	3
Rockets and Artillery	0.9	0.6	0.15
Aircraft	4.6	1.5	0.9
Command and control headquarters	3	1.5	0.9
Missile sites	7.6	4.6	0.6
Ship vehicles	1.5	0.6	0.3
Ports and harbours	30.5	15	6
Railway yards	30.5	15	6
Urban areas	60	30.5	3

When we started using aircrafts as a platform to observe the Earth's surface it was found useful as a reconnaissance tool during war and as a tool for carrying out topographical mapping as well as interpretation of natural resources of various kinds. In the same way, satellite remote sensing is used both for reconnaissance purposes as well as for the interpretation of natural resources but its use for preparing normal topographical maps is still to he developed fully, particularly on a larger scale.

Satellite reconnaissance systems like Big Bird (US) and Cosmos (USSR) equipped with photographic sensors, are well known. These provide data with ground resolutions from about 0.5 to 3 m, about once a day (Plate 16.1).

Civilian satellites lack much of the reconnaissance capabilities because of their limited spatial resolution, especially to observe a particular part of the Earth at a specific time. A number of advanced countries have launched satellites during the past, specifically for gathering military intelligence. They have also sold some of the data. A ground resolution of 2 to 5 m would be necessary for area surveillance purposes and a resolution between 0.5 and 2 metres is necessary for a closer look for detailed military activities and also for monitoring arms control.

The French SPOT satellite giving a spatial resolution of 10 m in black and white has been found to be useful to some extent for strategic reconnaissance. In addition, SPOT's capability to give overlapping pictures of the same area has been used for stereoscopic studies and for providing three-dimensional information of the terrain and generating a Digital Terrain Model (DEM). The DEM can be used to get perspective views draped with the actual picture of the terrain, which is very useful for strategic planning, as well as creating a GIS in conjunction with a strong positionally accurate cartographic database.

Defence Applications

These could be for tactical, strategic or for developing cartographic and geographic information systems. Tactical information may need very high resolution data, which have recently become available from Russian sources. They may become available from American sources too. For strategic applications the present available satellite data can meet most requirements, such as location of major industries, railway marshalling yards, airbases, dockyards, monitoring construction activities in areas not accessible to us, monitoring water features, trafficability based on terrain conditions targeting information, alert for surprise attack and monitoring military manoeuvres to some extent. All these are non-aggressive in nature and only aid in multiplying over a period of time, the military and weapons capability.

Most characteristics of the military satellites and their detailed use and incorporation in military organisations and weapons systems are classified information, so we cannot assess the degree of dependence on such systems. However, it is considered that major powers in the world do have dependable satellite intelligence systems for the purposes of verification, targeting and intelligence gathering. During hostilities, even though satellites may not be easily susceptible for interception, their support systems, like launching sites and ground stations, could be targets for attack. Therefore, the question of survival of and dependence on space activities will, to a great extent, be related to the particular case of warlike activity.

The type of satellite data needed in the case of military activity will also depend upon the temporal characteristics of the satellite, as the type of information may change very slowly or not change at all, such as geodetic or topographic data for which frequent observation may not be required. Varying information may need varying frequencies of observation. For example, observing military units and their movements, nuclear explosions, missiles, etc. may need very high observational frequency. In the case of obtaining positional accuracy of such data also, observational parameters have to be specifically designed. Parameters such as positional accuracy needed can vary from a few centimetres to hundreds of metres of the objects that are observed. For all such varying needs, different sensors have to be designed, which could be satellite-based in most cases or aircraft-based. Aircraft sensing may not always be possible as it is risky, particularly in a hostile environment. Since a photographic camera (with film) gives the best resolution, the film is often used in reconnaissance satellites as of now, but electronic, or electro-optical sensors are fast catching up.

The US Big Bird satellite was mainly used for area surveillance and used photographic film developed on board the satellite, converted to digital signals and transmitted to ground receiving stations. Probably they too used electro-optical sensors giving digital data, whereas, the US KH-11, is believed to perform close-look functions. The USSR used both in its cosmos satellites, giving high-resolution data. The photographic sensor was used to detect anti-ballistic missiles, inter-continental ballistic missiles deployment, military facilities, large scale troop movements, etc. Similarly, other sensors were used such as infrared thermal scanners, and radiometers to detect many other defence activities. According to a reconnaissance handbook published by the McDonnell Douglas Corporation of the US and reproduced by Dr. Jasani in 1982, the resolutions required in metres for interpretation tasks are given in the table.

Detection requires creating a class of units, object or activity of military interest. General identification requires determination of general target type. Precise identification requires discrimination within targets of known types.

At present, we are aware of geodetic or global positioning system satellites, meteorological satellites, land observation satellites and oceanographic satellites—each serving different and sometimes complementary functions. This only calls for effective camouflage measures.

Media agencies are also interested in obtaining high-resolution data from different sources to gather intelligence, although they have been using imageries from commercial satellites liberally for active reporting. One of the first news application of satellite data was regarding the Chernobyl nuclear disaster in April 1986. We could see the nuclear power plant from the Landsat 5 thematic mapper when it covered the area on April 29, 1986 (See Plate 16.2).

Information from the mid-infrared band 7 (2.08 to 2.35 μm (micrometer)) showed a bright area that was too strong a reflection than from the Sun. Data from the far-infrared band 6 (10.4 to 12.5 μm) showed that the cooling pond had a uniform temperature and was not being used for cooling, indicating that the reactors had been closed down for an emergency. The French SPOT satellite data also confirmed the above through its 10 m resolution panchromatic images. (Source: *Earth Observation,* December 1992, p 4). While space activities are adding value to the intelligence system, not every country may have the capacity and resources to take full advantage to enhance their capabilities. However, the time is not too far off when most major nations will ensure that their intelligence system is optimised, though more as a deterrent, by the liberal use of space imagery.

17
Mapping Wastelands through Remote Sensing

Wastelands can be mapped through remote sensing and thus we can take informed decisions on undertaking village-level development activities related to wastelands.

The total geographical area of India is about 329 million ha. Over a period of time, both geological and geomorphological processes resulted in the formation of a variety of land systems, such as mountains, hills, valleys, plateaus and plains. These different types of land are subjected to different types of utilisation to the maximum possible extent. Owing to the increasing pressure of population and finite land area, there is excess pressure on land, both for agricultural and non-agricultural activities. This has resulted in over-exploitation, certain parts of the land becoming unproductive, which become wastelands. In our country, the per capita availability of land declined from 0.9 ha in 1951 to 0.5 ha in 1980–81. Similarly, the per capita availability of cultivable land declined from 0.48 ha in 1951 to 0.25 ha in 1981. By 2000 it is likely to decline further to 0.15 ha. It is therefore essential that we make all efforts to identify and bring more land area into use for productive purposes.

Wastelands can be classified into three broad categories, such as,

1) lands officially classified as forest that have been degraded,

2) those under private ownership, not used

3) public land owned either by the Revenue Department, a government agency or village community.

In a broadcast to the nation, on January 1985, the Prime Minister said that a continuing deforestation has brought us face-to-face with a major ecological and socio-economic crisis. This trend must be halted. He proposed to immediately set up a National Wasteland Development Board with the objective of bringing 5 million ha. of land every year under fuel-wood and fodder plantations. He suggested developing a people's movement for afforestation.

The question of identifying wastelands and categorising them, over the whole country, was a major issue before any firm programme could be undertaken, as there were several conflicting and contrary estimates of wastelands, without any standard classification. After several discussions and consultations amongst several departments, an acceptable description of wasteland was evolved as follows.

"Wasteland for the purpose of the above means degraded land which can be brought under vegetative cover with reasonable effort, and which is currently under-utilised and land which is deteriorating for lack of proper water and soil management or on account of natural causes. Wasteland can result from inherent/imposed disabilities, such as by location, environment, chemical and physical properties of the soil of financial or management constraints" (NWDB-1987).

A 13-fold classification of wastelands was also evolved such as ; 1) Gullies and or ravines, 2) Upland with or without scrub, 3) Waterlogged and marshy land, 4) Land affected by salinity/alkalinity/coastal or inland 5) Shifting cultivation area, 6) Sands–desert/coastal, 7) Mining/Industrial wastelands, 8) Under-utilised and degraded notified forest land, 9) Degraded pastures/ grazing land, 10) Degraded land under plantation crop, 11) Barren ,cky/ stony waste/sheet-rock areas, 12) Steep sloping area and 13) Snow ɔvered and/or glacial area.

Detailed descriptions of each of the above categories were alsu provided by an expert group constituted by the wasteland development board in consultation with the NRSA.

Till 1985, no serious attempts were made to systematically prepare a consolidated map on any scale, showing the geographical location and spatial distribution of different types of wastelands in the country. The National Commission on Agriculture in 1976 made an estimate of wasteland in India, which then was revised by the soil and water conservation division of the Ministry of Agriculture (1982). The official estimate of wasteland in India which is more than half of the country's geographical area (Mukherjee *et al.* 1985) was nearly 175 million ha. Subsequently, it was revised to 148 million ha. Similarly, there were several pronouncements till the NRSA rapidly covered the entire country by interpreting the satellite remote sensed data on a 1:1 million scale and estimated a rough figure of 53.3 million ha. (See Fig. 17.1)

This satellite imagery was already obtained a few years before for mapping and estimating forest cover in India; attempts were made again. This estimate had its limitations, due to the very small-scale of mapping and lack of intensive field checks and the rapidity which this work was carried out. However, these maps helped in identifying areas and districts in India, which had somewhere around 15 to 20 percent of the area covered by some kind of wastelands. It was at this time, that the US Landsat Satellite carried on board a sensor called Thematic mapper, which provided a ground resolution of remote sensed data at 30 m (i.e. objects of size approximately 30 m across can be identified, theoretically).

As a result of several discussions and consultations, a methodology was developed to identify wastelands in India and map them on a scale of 1:50 000 to cover about 146 districts (supposed to contain about 15 percent with wasteland) at the first two phases of the operation. Accordingly, a detailed project proposal was prepared and presented to the wasteland development board, which got approved almost immediately, as there was no other method of achieving this task rapidly. A time schedule of 12 to 14 months was set, a sum of around Rs 3 crores was earmarked and the work was entrusted to the NRSA, which was identified as the national co-ordinator. This directly reported to a committee constituted with the chairman, NWDB, the chairman, ISRO, the Surveyor General of India and the Secretary NWDB acting as member secretary to guide and monitor the work.

Approximately 1 300 satellite images on 1:50 000 enlargements were prepared, about 2 800 base maps were traced out of the 1:50 000 topographical maps on which wasteland categories identified on the recent Landsat remote sensed imagery was superimposed. The additional feature was to include village boundaries and tahsil boundaries, also superimposed as best as possible, within the limitations imposed due to the different scales of such maps not fitting within the national mapping framework accurately.

A total of 2 813 maps resulted, completed in about 14 months and distributed to different states for use, with a limited number of copies only, as large-scale printing was not resorted to, owing to financial and time constraints. A booklet was prepared to explain how these maps could be used on the ground and distributed widely.

This exercise was perhaps the first of its kind in India and resulted in the following:

i) A number of scientists were trained in interpreting different categories of wasteland intensively and rapidly, in about 2 months, as there was acute shortage of personnel.

ii) The work got distributed to about 20 work centres, located in different parts of the country, generally comprising state remote sensing application centres which were coming up slowly, the All India Soil and Land Use Survey, the National Bureau of Soil Survey and Land Use Planning, the Centre for

Studies on Resources Engineering in IIT, Bombay, the Maharashtra engineering Research Institute, Nasik, and the various centres of the Department of Space, like the Space Application Centre, the NRSA, the Indian Institute of Remote Sensing and the Institute of Remote Sensing, Anna University, etc., besides the Survey of India which was the strongest pillar of support by active participation. Over a 100 scientists and about 200 cartographers were employed, spread all over India.

iii) This project provided a major confidence boost to the scientists involved in Remote Sensing Applications of the DOS/NRSA, etc. that they could rapidly carry out such surveys of national requirement at short notice and produce almost near real-time information for sound planning and also periodically monitoring changes that occurs and also estimate progress achieved with respect to investment made.

iv) It provided first-hand and reasonably accurate locational details of waste-lands at almost village level to enable undertaking development activities related to wastelands.

A sample picture of the actual satellite data (IRS–LISS II) and another picture of the resultant map produced showing wasteland is included here which covers part of Gwalior in Madhya Pradesh covered by Survey of India topographical 1:50 000 scale map sheet No. 54J/4. The area of different categories of wasteland was calculated as follows, failing in this map sheet.

1) Gullies or Ravines – 4174.00 ha.

2) Upland with or without Scrub – 7086.00 ha.

3) Under-utilised/degraded notified land – 6386.45 ha.

4) Barren rocky/Stony waste/Sheet-rock area – 157.50 ha.

This particular map sheet was produced by the remote sensing centre of the Jawaharlal Nehru Technological University (JNTU) recently. There was no other category of wasteland found in this mapsheet area.

The remote sensing application activities have, in the last decade or so, spread to many departments, institutions and also to academic institutions and universities. While some universities have introduced courses for academic studies only, there are others who have also undertaken productive work on payment and on an operational basis.

The JNTU, Hyderabad, took advantage of the NRSA located close by and has also entered into a Memorandum of Understanding with the Survey of

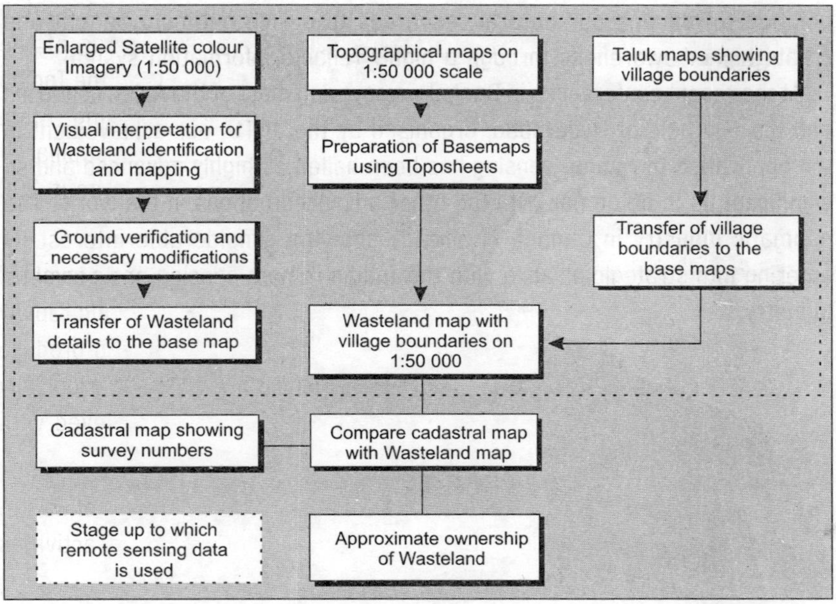

Fig. 17.1 *Flow diagram showing the methodology adopted in mapping wastelands using remote sensing data*

India for conducting advanced academic programmes at the Master's level.

The progress made by them during the past four years, covering activities connected with remote sensing, mapping, etc. is noteworthy. There are some other universities also in our country, taking to space application activities seriously.

It can thus be concluded that but for the availability of such high technology tools like space remote sensing, both in digital format and as well as in visual or analogue format, it would not have been possible to identify and map wastelands in our country so rapidly, to provide almost near real-time information. A third phase covering another 85 districts appears to have been taken up and completed recently, under the leadership of NRSA. It may not be enough if only identification and mapping of wastelands is carried out, but it is essential to make full use of them to economically bring such wastelands in to productive use to meet the growing demand for food, fodder and fuel.

Now that India has its own remote sensing satellites, with a ground resolution of 35.5 m and the possible IRS-1C with a ground resolution of better than 10 m launched in the third quarter of 1995, we should be in a better position to manage our natural resources to derive optimum benefit and create higher awareness through a highly reliable information system.

At the recent conferences in Bangalore covering most of the Asian countries and the one held at Hyderabad, organised by the JNTU, our achievement in the application of remote sensing has been hailed as highly advanced and so significant as to be on par with the other advanced nations in the world. The geomatic industry in Canada is already showing considerable interest by entering into strategic alliance with the Indian remote sensing and geomatic industry.

18

Digital Forest Management

Satellite data has a direct bearing on Forest Management, in various applications like inventorying, checking forest fires, biomass and fuel-wood assessments, both on a state and national level.

I shall now try to explain how Remote Sensing and GIS can be used to aid in various aspects of management of our forests more effectively and based on near real-time information. Foresters have to take many decisions, sometimes short-term and sometimes long-term, and GIS has been found to be effective—it is essentially computer based.

While remote sensing is generally talked about from satellite platforms, obtaining data from the aerial platform cannot be excluded, as a combination or integrating information derived from satellite altitudes as well as aerial altitudes, which combined with ground information will enable a better under standing for arriving at various decisions in the area of forestry management. As of today, in our country an effective and near real-time and modern forest information system does not seem to exist, actually management has become more problematic.

In India, the Central Government exercised control on formulation and regulation of national policies on the subject of forest management, but the respective State Government forest department is fully responsible for managing the forests falling within its jurisdiction. Generally, almost all forest departments of the State Governments are organised similarly, keeping within the general administrative pattern. While in the past forest management essentially aimed at generating revenue through production, the emphasis is now more on forest ecosystem management for maintaining a sound ecological and environmental stability.

Forests as a major source of energy was focussed upon sometime in 1973, worldwide, when the oil crisis emerged. Remote sensing was a major tool to collect urgently needed data related to monitoring changes in forest cover, assessing land use and forest land degradation, evaluating the productivity of the land and providing information not only for forest inventory but also for direct inputs to forest management and strategic planning. While it was recognised that a forest cover of around 33 percent is desirable for maintaining environmental stability and ecological security, it could not be achieved, as the forest was rapidly dwindling. Also there was no authentic mechanism to bring out the status of forest cover, on a near real-time basis, until spaceborne remote sensing emerged as an operational tool, the data obtained being free from any subjective bias.

By 1983 or so it was emphatically brought out that in India our forests were reduced by about 2.5 to 3 percent during the decade 1972–1982, by a comparative forest cover study carried by the NRSA, using remotely sensed data obtained from Landsat (US) satellite, and that the closed forest covered by that time was only around 12 percent of the total land cover. Since then, the Forest Survey of India has been carrying out a biennial monitoring of forest cover using satellite remote sensed data on a scale of 1:250 000 (See Plate 18.1).

Some of the main information the forest managers would like to have on a near real-time basis would be

1) dynamic changes and surveillance of forest cover,
2) forest type distribution,
3) plantation monitoring,
4) estimation of forest growing stock,
5) forest inventory and volume estimation,
6) preparation of stock tables and yield calculations,
7) preparation of zones or treatment areas,
8) assessment of ecological and biodiversity related issues,
9) forest change and conversion studies,
10) forest damage assessment due to forest fires,
11) grassland identification and productivity assessment
12) biomass and fuel-wood assessments, and developing other related information systems, which could be used for effective management practices.

With the advent of rapid advancements in computer technology, and with the use of GIS techniques applied carefully, one could develop an up-to-date, dynamic digital information system, to not only keep abreast of what is happening at a frequent intervals of time, but also to identify thrust areas on forecasting and prediction models based essentially on remote sensing data supported by other collateral data. It may also be stated here, that while information may be required from a national and state level, we will also be needing information at district level and even at micro-level and it may be necessary to use both satellite based remote sensing and aerial data, supported by appropriate ground truth measurements.

The present scenario in the use of remote sensing in forest studies, indicates that for forest cover monitoring or surveillance, the use of this technology at the national and state level and mapping on a scale of 1:25 000 is possible. This is operationally being done by the Forest Survey of India, to produce and report information on a biennial basis. Such information has been found to be quite reliable and can be generated within a short span of time and, if necessary, in digital format. Besides this, the study of spatial distribution of forest types could also be necessary and useful, particularly to provide a better insight to the working plan officials as to what type of species need to be introduced either by replacement of natural forest or regeneration of degraded forests in the area in a more scientific manner. Preparation of forest type maps and their distribution at the block level on a scale of 1:50 000, is now possible using satellite remote sensed data to aid forest managers.

Agencies such as the NRSA, Space Application Centre of ISRO, etc. have, in co-operation with the Forest Administration in the country carried out several studies using remote sensed data obtained from different satellites like the Landsat (USA), Solyut-7 (Russia), SPOT (French), IRS 1A & 1B, etc., in the area of mapping forest types, density and stock mapping, monitoring deforestation and afforestation, monitoring encroachment in forest areas, mapping forest banks, assessment of production potentials, and wildlife studies; all of which have an important bearing on managing forests.

The use of satellite remote sensing data has amply demonstrated the feasibility to identity forest-fire damaged areas. The studies carried out during 1991, in parts of the Nagarhole wildlife sanctuary, facilitated accounting the damages to an area of about 68 sq km. Recently in March 1994, in the Simlipal reserve forests, the extent of fire damage was assessed using the Indian remote sensing satellite (IRS) 1-B and estimated the fire damage to an area of 600 sq km against the 200 sq km reported. The use of periodic seasonal data of certain specific forest areas prone to forest fires could be used by the forest managers to enable taking certain fire prevention measures or mounting fire operation measures in advance to mitigate possible damages.

In the recent Asian Conference on Remote Sensing, held in Bangalore, in November 17–24, 1994, the delegates from Thailand had reported how they have effectively used satellite remote sensed data.

It has also been pointed out that an effective forest management information could be operationalised using the GIS techniques by integrating information obtained from different sources. This will facilitate organisation of a database, data updating, retrieval of relevant data when required and its analysis, possibly at district level. In fact, a digital data information system for forest could be a major input for effective management of forests. Towards this end, the Department of Space is already engaged in the development of a National Natural Resources Information System. This also visualises an elaborate information system encompassing various aspects of forest management with a holistic approach.

Satellite remote sensed data has the advantage of being digital in character, enabling the use of computers, viewing large areas in one scene, and obtainable at regular intervals of time covering same area. However, as of now, the spatial resolution now available limits the observation and mapping to a level of around 1:50 000 scale, although in certain cases, mapping at 1:25 000 has also been possible. By the turn of the century, satellite remote sensing data is expected to reach a much better quality and provide much higher spatial, spectral and temporal resolution data providing mapping capabilities to a larger scale, than at present. Already there are reports from the US about various large private companies vying to launch satellites providing data of 1 to 2 m spatial resolution. While estimating volumes and providing large-scale three-dimensional stereoviewing capabilities, it still remains mainly the domain of aerial photography. Technology innovations may lead satellite data products to match aerial photography standards, at least in the range of small- and medium-scales of mapping.

In conclusion, it must be emphasised here that remote sensing has come to stay, and its usability in managing our forests on a sustainable basis cannot be overemphasised. Conservation and restoration processes especially with regard to a renewable natural resource like forests, is essential to contribute towards environmental stability (Plate 18.2). For achieving this near real-time information some system is absolutely necessary. It is, therefore, suggested that all our forest departments in the states should soon get to know more about modern techniques such as remote sensing and GIS, so that our forest management practices could keep pace with our people's growing needs. The earlier we make a beginning in this direction, the better it would be for ourselves. Political bosses and bureaucrats should not consider the expenditure incurred in these directions as mere expenses, but as investments being made for long-term benefits. I would like to mention here that some of the technical presentations and articles contained in the

15th Asian Remote Sensing conference here at Bangalore and the Indian Geography Congress held at Hyderabad have provided valuable inputs in writing this article.

19

Remote Sensing: Sky's the Limit

A survey of all the major space programmes in the world is undertaken here. Against this background, the Indian space programme is described and found to be on part with the best in the world.

I have dealt with various aspects of satellite remote sensing and its applications. Perhaps, it is now time to review various satellite systems, relevant to remote sensing in the recent past, the present and future, not only in India but also on a worldwide basis. During the past 35 years or so, it has been reported that about 50 satellites relevant to remote sensing have been launched the world over, and in the next 15 years, a similar number of satellites is expected to be launched by different countries. While an enormous volume of data about our planet has been collected through this process, the utilisation has not been much, say, just ten percent.

One may wonder why this is so. Dr Macdonald, president of Macdonald Detwiler of Canada, in one of his talks identified two types of users of space remote sensed data; one, the scientific users and the other, the operational users or commercial users. He pointed out that while the scientific users develop and produce knowledge, the operational users produce wealth. While for the global environmental scenario, scientific inputs are required, for the fast changing and competitive economic scenario, we require an operational or commercial approach. More efficient planning and management of resources therefore becomes a necessity.

Three major application areas of space technology are: space communication, meteorology and remote sensing of the Earth and its resources. We shall concentrate on remote sensing-related development in this article.

The series of satellites now known as LANDSAT (Land Space Application Technology) launched by the US, evolved in concept from the photographic observations of the early Mercury and Gemini orbital flights. Data from those manned Earth orbital flights indicated the practicality of observing from space orbit what is broadly referred to as "earth resources." These observations and the thoughts they generated led to the NASA (National Aeronautic and Space Agency of the US) satellite programme that developed the Earth Resources

Technology Satellite-I (ERTS-I), which was launched in July 1972, and which was later renamed Landsat 1.

Since it was found necessary to have a non-photographic sensor system with sequentially acquired images in a specific framed format, a television type of system called the Return Bean Vidicon (RBV), which could acquire framed images sequentially and which could be telemetered to ground station was devised. However, the user interest revolved around geology, cartography, forestry, agriculture, etc. and made technologists look for a system, which could meet multi-spectral requirements more efficiently. It finally resulted in the development and use of multi-spectral scanners along with RBV, mainly focussing on the visible bands of the electromagnetic spectrum (EM Spectrum) in the region of 0.5 to 0.6, 0.6 to 0.7, 0.7 to 0.8 and 0.8 to 1.1 micrometers.

This was found to be an excellent decision, when data was actually received from Landsat 1 later on, and thus an operational Remote Sensing Satellite System emerged. The nominal altitude of this satellite was 913 km, with an 18 day repetitive cycle, with a sun-synchronous orbit, crossing the equator at about 9.30 a.m. local time, which gave a uniform Sun angle illumination. The ground coverage in each frame being 185×185 km, it provided a synoptic view, covering a very large area of the Earth's surface.

Landsat 1 had two sensors, one called the Return Beam Vidicon Camera (RBV) and multi-spectral scanner (MSS). The RBV observed in the spectral range of Band 0.460–0.600 micrometer (green region); Band 2 0.569-0680 micrometer (red); 0.660–0.80 micrometer (Band 3, near infrared). The MSS which is a line scanning device with an oscillating mirror, produced four synchronous images at Band 4, (green) 0. 5 to 0.6, micrometer Band 5, (Red) 0. 6 to 0. 7 micrometer, Band 6, lower infrared 0. 7 to 0. 8 micrometer. The data observed was relayed to the Earth at places where receiving stations were located, as the one at Shadnagar near Hyderabad in India.

Landsat 2, with a similar sensor configuration was launched on January 22, 1975, for continuity of data. Subsequently on March 5, 1978, Landsat 3 was launched, again with a similar sensor configuration. However, Landsat 3 had an additional band responding to emitted thermal infrared radiation in the range of 10.5 to 12.4 micrometers of the EM spectrum; but they did not work for long. All three satellites were retired after about five to eight years of service. All three satellites provided an approximate ground resolution of 80 m by the MSS imagery. The data received from the RBV camera did not find favour among the users due to various reasons, and was discontinued.

Landsat 4 and 5: These two satellites with improved spatial, spectral and temporal resolutions formed the bridge between the old and the new in the

Earth resources satellite system. These were launched on July 16, 1982 and March 1, 1984, respectively, with near-identical sensor characteristics. This sensor called Thematic Mapper (TM) provided a ground resolution of 30 m, with a seven channel data which included a thermal infrared channel also (10.40 to 12.50 micrometer of the EM spectrum). It provided a revisit every sixteenth day, orbiting at an altitude of about 705 km and a ground coverage in each scene of 185 x 185 km. These two satellite systems provided a tremendous improvement in our understanding of the Earth's resources. Landsat 5 is still in orbit but it rarely provides data. Landsat 6, was planned with some improvements over the Landsat 5, but the launch was a failure about two years ago and a replacement in the form of Landsat 7 is scheduled to be launched by NASA in the near future.

France launched a satellite called SPOT (Systematic Probatoir d'Observation De la Terra) in February 1986, which provided a new generation of remote sensing satellite systems, with a quantum jump in spatial resolution to 10 m, and also capable of providing stereo images. It orbited at an altitude of about 833 km with a revisit capability of 26 days. It had a multi-spectral scanner capable of giving data at a 20 m spatial resolution and a separate sensor system providing data at 10 m ground resolution, but only in a black and white panchromatic mode.

The multi-spectral bands provided data on three channels only, but was considered adequate for most Earth resources application and studies. However, the data format or ground swath in each scene was only 60 km, which some consider as an advantage, but others as uneconomical. However, the improved spatial resolution and stereo capability took the user community to a new area of opportunity, including its use in cartographic map updating, and also in very special cases data sets at 1:25 000 scale.

It provided an opportunity to obtain Digital Elevation Model (DEM) in several parts of the world and gave a boost to GIS, the world over. We in India received SPOT data directly for a few years, but discontinued it for various reasons. This satellite has an on-board tape-recorder on which data pertaining to India could be recorded and received by downloading in France or at any other place. SPOT 2 and 3 were launched in 1990 with identical payloads as SPOT 1. They have programmes to launch a modified SPOT 4 in due course.

The Japanese also launched satellites called MOS (marine observation satellite) and JERS (Japanese Earth resources satellite). Data from these are not available in India.

ERS (European Remote Sensing Satellite): It was already known that cloud cover was a limitation in obtaining remote sensed data using the existing optical sensors, carried by satellites. To overcome these limitations as well as to obtain more information, the European Space Agency launched a microwave satellite called the ERS.

Having outlined some of the major remote sensing satellite systems belonging to foreign countries, let us now see the picture in India as the Indian effort has been considered to be one of the best in the world, on par with the developed countries. We have established our capability to design, fabricate and even launch our own remote sensing satellites and recently data from our satellites are being received, processed, stored and distributed by the US to the world community.

A recent agreement entered into by the Chairman of ISRO with his counterpart in the US, specified their continued interest in receiving Indian satellite data for the next 10 years or so, on a commercial basis, and it is reported that the economic benefits would be of the order of one billion dollars or so. The Indian space programme envisages between now and 2005, about eight satellites planned for launch.

The Government of India set up the Space Commission in 1972 to promote the development and application of space technology and space sciences for the socio-economic benefit of the nation. The Department of Space (DOS) is responsible for the execution of activities in the country's space applications, space technology and space sciences. The Indian Space Research Organisation(ISRO) was formed simultaneously taking up activities in launch vehicles, design and fabrication of satellites, development of launch capabilities and application of space technology in all its aspects such as communication, meteorology and remote sensing.

The National Remote Sensing Agency was established in 1975 in Hyderabad as an autonomous body, to develop utilisation and operational capability of remotely sensed data for the survey and management of natural resources of the country as well as to receive such data directly from orbiting satellites, process them, store and undertake distribution.

Aryabhata was the first Indian experimental satellite that was successfully launched in April 1975 in an approximate 600 km orbit above the Earth, with the help of the Soviet intercosmos rocket. The orbit was inclined at about 50° to the polar axis. This provided confidence in indigenous design to fabricate space-worthy satellites, and also to evolve the methodology of conducting a series of complex operations on the satellite in its orbital phase. It also provided the opportunity to set up ground-based receiving, transmitting and tracking systems. This satellite functioned for almost six years.

Next was the Bhaskara-I satellite, launched with Soviet help on June 7, 1979, mainly as an Earth observation experiment to collect, process, analyse and disseminate data on hydrology, forestry and geology using two television cameras. Its life was for one year. Bhaskara II was launched on November 20, 1981, with Soviet assistance, which transmitted pictures of frames covering 341 x 341 km of the surface of the Earth and with a spatial resolution of about 1 km and orbiting at 525 km. The data obtained was on two spectral bands of 0.54 to 0.66 micrometers and 0.75 to 0.85 micrometers. At the end of two years, this satellite had completed 11 000 orbits around the Earth. The closing of the Bhaskara II mission in 1984 marked the end of the experimental phase of ISRO, and its entry into an operational phase.

The first operational Indian Remote Sensing Satellite (IRS-1A) was launched on March 17, 1988, which is the largest and the most complex satellite developed by Indian Space scientists. It was launched from the Baikonur cosmodrome in the USSR. It was finally put into an orbit at 904 km above the Earth. It weighed 975 kg and was planned for a three-year-lifespan, although it served more than double this period. It carried two pushbroom type of cameras, one called the linear imaging self-scanning sensor (LISS I) with 72.5 m ground resolution and a swath covering 148 km of the Earth's surface.

The second set of two cameras called LISS II A and LISS II B provided 36.25 m ground resolution and a ground coverage of 74 km each. Both put together covered about 148 km allowing a 3 km overlap. They provided data on four bands called band 1, 2, 3 and 4, observing at 0.45–0.52 micrometres, 0.52 to 0.59 micrometres, 0.62 to 0.68 micrometers, and 0.77 to 0.86 micrometers. Its revisit was once in 22 days. This IRS-1B, was launched in 1991, to provide back up to the IRS-1A, they carried the identical payload as that of IRS-1A. IRS-1A and 1-B, when combined, revisit the same area every 11 days.

The recent (October 1994) spectacular launch of the Polar Satellite Launch Vehicle (PSLV), which placed the indigenously developed Indian Remote Sensing Satellite-P2 (IRS P2) into orbit, bears testimony to the technological progress made by India towards providing reliable and sustained operational space services. This satellite, IRS P2 orbits at 817 km with a revisit capability of once in 24 days, providing a ground resolution of about 33 to 37 metres in four spectral bands.

The second generation remote sensing satellites IRS-1C and 1D, with improved spatial resolution and different types of sensors with stereo imaging capability, is due to be launched in the third quarter of this year. It will take us, if not ahead of most countries, at least on par with the best in the world. This satellite is expected to have an on-board tape-recorder, so that data from any part of the world can be acquired, recorded and played back to download anywhere (Plate 19.1).

The most important improvements are that it can have a revisit capability of five days, a high resolution panchromatic camera system giving a ground resolution better that 7 m and an additional sensor called (WIFS) wide field camera, with a ground swath of 770 km with a resolution of 188 m. It will also carry an LISS III camera, giving a ground resolution of 23 m in four bands which can be considered as an improved version of LISS II carried by IRS 1A and 1B.

Every move our space department has so far taken has been extraordinarily impressive and rewarding. By the turn of the century, it may not be surprising if our satellite remote sensing data becomes the most coveted in the world (Plate 19.2). There are several more remote sensing satellites that are being planned by several countries such as the US (Landsat 7), Canada (Radarsat-1996), France (SPOT Series), Japan, the European space community, Brazil, possibly China and Russia too, each with specific improved features. All these will flood the world with an enormous volume of data. Unless private enterprises use this data and add value to them to meet specific user needs, much of it may remain in the archives. It is, therefore, necessary to create conditions even now to enable use of this data for the economic benefit of our people.

20

The Impact of Multi-Purpose River Projects

An analysis of the application of remote sensing for studying long-term environmental and ecological impacts of multi-purpose river valley projects.

In the past two decades the general awareness of environmental issues has gained considerable importance and every large-scale project has been subjected to some kind of environmental impact study and scrutiny. With the availability of remotely sensed data, large areas of the Earth's surface can be viewed at frequent intervals of time. Such a synoptic view will make available information on a near real-time basis. It also makes for a realistic consideration of the environment and ecological impact studies, monitors changes and activates early remedial measures. While one cannot easily brush aside the views expressed by environmentalists, one has to look into the beneficial aspects of such large projects from the requirement point of view, by accepting certain standards to be followed.

No country can survive without exploiting its natural resources and developing its economy. The development of water resources for irrigation and extended agriculture, and for generating power by harnessing rivers through dams are essential considerations. The 1987 Brundtland Bulletin on 'our common future' states that 'sustainable development is about using today's environmental resources without exhausting them to the disadvantage of future generations.'

Several large-scale water resource projects took a long time to complete and involved enormous expenditure. These irrigation and power projects have changed the face of the country's economy in many ways. Perhaps some of the undesirable or avoidable adverse side-effects were not seriously thought of at that time, like waterlogging and salinity caused by poor drainage systems, sedimentation of reservoirs, submergence of forest land, displacement of persons living in submerged areas, change in land use pattern affecting the ecological stability and posing a threat to biodiversity.

Studies on environmental impact were carried out using remote sensing as a tool. One example is the study done by the NRSA, under a project initiated by the Department of Environment (DOE), Government of India, covering the Idukki area in Kerala.

The studies were undertaken between 1980 and 1983. While the DOE attempted to integrate studies from various angles carried out by various national and professional agencies, the role of the NRSA was only to point out the changes that have occurred over a fairly long time period, using remote sensing as a tool, and also to obtain baseline information in the form of thematic maps, on which the detailed findings of other studies could be incorporated to infer the inter-relationships.

A multi-purpose river valley project, like the one in Idukki in Kerala, is likely to cause appreciable ecological disturbances to the surroundings. If we understand the causes for such disturbances, we would be able to find remedial measures at least to some extent. This requires data and information on a near real-time mode covering large areas. Remote sensed data is made use of for meeting this requirement, which cannot be done by any other known means at present. In this project, while satellite remote sensed data was used with respect to natural resources at a reconnaissance level and to monitor changes that occurred over an approximate 10 year period, aerial photographic and aerial multi-spectral scanner surveys were also carried out to study thermal changes and other aspects.

Here, we shall only discuss the various studies that were carried out and maps prepared using satellite based remote sensed data. Essentially, the information generated covered land use/land cover and soils, geomorphology, drainage, water features, etc. The studies were carried out twice, once with satellite data pertaining to 1973 (Plate 20.1), when the dam had not been constructed and again using satellite data of 1981 (Plate 20.2), when the construction was completed and water stored in the reservoir. This enabled a comparative study. The satellite data then available was from the Landsat 1, 2 and 3 satellites (US satellites) and the data had a ground resolution of about 80 m. The scale of the maps had to be kept to 1:250 000, in view of the limitations imposed by the special resolution of the satellite data available at that time.

The main objectives of the study as far as remote sensing data is concerned was as follows:

1) Prepare a broad land use, forest/vegetation cover and types, and soil classes map, using the Landsat data pertaining to Jan-Feb 1973 and 1980, covering the entire Periyar river basin.

2) Prepare a river/streams-course and drainage pattern map by visual interpretation of Landsat imagery pertaining to both the periods, and compare them to find changes.

3) Prepare a geomorphic map of the area to show macro-landforms, by visual interpretation of Landsat imageries.

4) Obtain aerial multi-special data and aerial photographs by flying over an area of about a 1 000 sq km surrounding the river basin. This data enabled more detailed studies of the area.

While there was loss of natural forest covering large areas, and which were replaced by plantations, it cannot firmly be stated that it was all due to the hydroelectric project. The direct impact of the project to the immediate vicinity of the project site can be because of the following.

1) The submergence under water: a considerable area of grassland or rock outcrop was lost due to inundation of water. There were some patches of natural forest which also disappeared.

2) The natural vegetation in the immediate neighbourhood was destroyed to some extent and altered due to clearing of areas for establishment of project offices and other infrastructural facilities.

3) The affected people had to be rehabilitated in other nearby areas which caused clearing of some amount of natural forest.

Since the use of satellite data at that time gave a more qualitative assessment, it was necessary to study the aerial photographs and scanner data which was obtained in 1983, covering approximately 1 000 sq km to make a more detailed assessment.

However, such a study could not be carried out by reconstructing the situation as it existed in 1972-73, because the aerial data of that period was not available. While in general terms, it can be stated that there were considerable changes in land use, which could be attributable to the construction of the hydroelectric project, a more detailed study of the flora and fauna will be necessary in future, to quantify the changes. However, by effective enforcement measures, further encroachment to the natural forest could be stopped to maintain ecological stability.

While a certain amount of impact cannot be totally avoided, a more detailed investigation in such large-scale projects could indicate areas where protective measures could be effected. For this, the most suitable data that can reveal the existing status of the local ecological situation is satellite based remote sensed data, the quality of which is considerably improved now and will certainly improve further in the coming years. A similar study was carried out in 1983 to cover the Silent Valley area, to acquire baseline information, as there was a proposal to build a dam to harness the water resources in that area for irrigation and drinking and to assess the extent to which the environment would be affected. The project was abandoned due to pressure from environmentalists, and it was declared a protected site. It could therefore be concluded that in all such large-scale water resource development projects a study of satellite remote sensed data, supplemented by aerial data, ought to be carried out.

21

What Space Technology has to Offer

Improved space technology has given us better quality and higher resolution satellite remote sensed data which can lead to optimisation of micro-level planning. A case-study is the Chamba District in Himachal Pradesh and 156 other districts adopted by the Planning Commission.

All development, particularly those activities related to land and water, are planned on information which in the past was not easily available, or in most cases outdated. Integrating various information to avoid any adverse impact was also not well understood earlier on.

This article highlights the concepts that have been evolved in consultation with the Department of Space, Government of India. Called the Integrated Mission for Sustainable Development (Fig. 21.1) the programme covers 157 districts as projected by the Planning Commission. Studies in 21 districts taken up in the first phase showed encouraging results, but how far these results are sustainable and to what extent they are successful is yet to be seen. The main purpose is to highlight how remote sensed data can contribute to planning for sustainable development and how far it can be implemented. Remote sensed data by itself cannot provide very detailed information for micro-level planning and development as yet.

In the recent past, particularly in developing countries, nothing has received closer attention than the concept of 'sustainability'. The Bruntland Commission states that development should be sustainable and sustainability means that humanity meets the needs of the present without compromising the ability of future generations to meet their own needs.

In another publication, Harvey Brooks says that world conservation strategy should include management of use of a resource so it can meet human demands of the present generations without decreasing the opportunities for the future.

As for agriculture, research organisations lay emphasis on agricultural systems and technologies that minimise environmental damages. There is also an emphasis on the need to use resources to promote long term productivity and growth along with the maintenance of environmental standards.

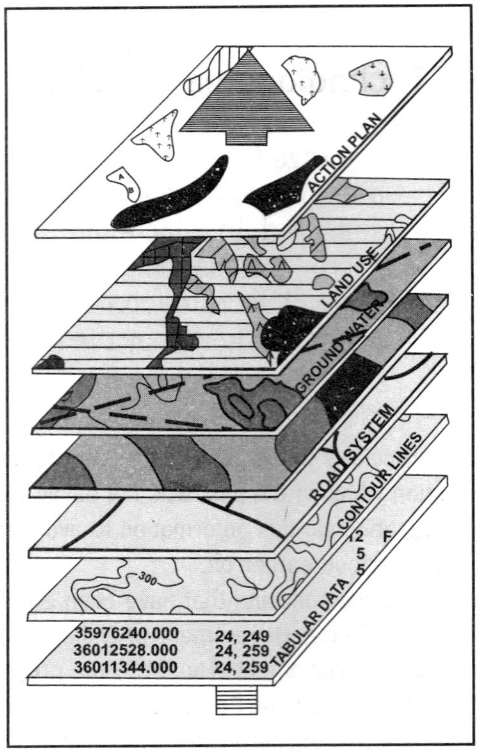

Fig. 21.1 *Integrated study to combat drought for sustainable development: Technical guidelines (Courtesy: NRSA)*

Since human needs keep changing, it is difficult to predict what the future holds and what steps can be taken to see that these needs are met satisfactorily. However it can be presumed that sustainability is a characteristic of a system and not a technology. Development implies, to some extent, growth which is quantitative in terminology. New frontiers in technologies now available are Biotechnology, Space technology, Information technology, Micro-electronics and Management technologies.

These technologies can be blended to apply to various situations. We shall talk about what space technology offers and how remote sensing as a tool can provide the most up-to-date information on natural and environmental resources. This could be integrated with other relevant information to provide a management strategy to plan alternative development paths towards sustainability concepts.

The main information required about land and water resources is obtained from recent satellite remote sensed data (from IRS 1A/1B satellites). This is integrated with various auxiliary and socio-economic data, to evolve alternative strategies for development planning. Attempts are being made to evolve site-specific solutions according to land capability assessment, as terrain conditions are diverse in our country. Each of the land capability classes is further divided into several micro-level units depending upon the soil limitations related to erosion, salinity/alkalinity, moisture-holding capacity, etc. for identifying suitable conservation/management practices for their development. In addition to carrying out studies in these 157 districts, 80 more districts/blocks have been identified for undertaking studies on a priority basis, based on special requests and decisions made at the highest level.

NRSA has prepared a manual describing the methods to be adopted. A draft has been circulated to a number of agencies. The main action involved in this study is the creation of a database of various natural resources: forests, soil, water resources (both surface and ground water) natural vegetation, land use/land cover, both in visual map form and perhaps in digital format as well.

The 1: 50 000 scale topographical map of the Survey of India is being used as the cartographic database on which other natural and socio-economic data can he superimposed to observe the effect of integrating relevant themes depending upon terrain conditions. A slope map has also been prepared based on the topographic maps on a 1: 50 000 scale. In preparing map information of the resources mentioned, care is taken to ensure accuracy and reliability through what is called ground truth observation or ground checks. Based on the generation and integration of such information, the production potential of the land is estimated, so that better agricultural productivity in areas suitable for horticulture, fodder, pasture, etc. can be planned. Wherever computerised information is generated, the use of a suitable GIS can be developed to support the various decisions that may have to be taken.

In this effort, several agencies are involved and the initial work itself is being assigned to various State Remote Sensing Application centres already in existence in addition to some of the Regional Remote Sensing Service Centres, under the Department of Space.

The concept of such integrated resource development has many advantages as these are now possible using satellite based remote sensed data, although the level, intensity and scale would be only of the order of 1: 50 000 scale, most suitable at district level.

Such a concept is not new, as can be seen from a publication called *Resources Analysis for Integrated Development* covering Sultanpur district in Uttar Pradesh, published by the Indian Agricultural Research Institute in 1986.

The major difference in what was thought of at that time and what is being undertaken now is that the information now available is almost near real-time and also amenable to computer processing. In the past, most of the information took time to obtain and to a certain extent, was not uniform. Also, the cost of obtaining all such data was perhaps higher. Continuous updating of natural resources information is much easier now using satellite remote sensing techniques than what was possible earlier.

In one of the studies covering the Jabua district of Madhya Pradesh, which had very low rainfall (about 600 mm average per annum), it was found that the productivity in the district was very low because of incorrect land use practices, inadequate fodder to feed a large cattle population and extensive forest degradation. The results of the studies were discussed with local officials and it led to an action plan involving the improvement of the water conservation measures and ground water recharging measures. This way land could be put to better use by developing silvipasture, agroforestry and agrohorticulture. Similar studies carried out in certain specific blocks of the Anantapur district resulted in improved ground water availability and farm practices yielding better income to the local population.

Another study covering some of the districts of Himachal Pradesh, by the Regional Remote Sensing Service Centre at Dehra Dun was mainly on soil conservation, slope protection and prevention of landslips, conservation and management of water resources, agricultural and horticultural development and fodder resources development. The aim was to draw up specific plans by integrating natural resources information with socio-economic data, which could meet the needs of the people and may lead to a sustainable development process. Five districts (Chamba, Lahane and Spiti, Kinnaur, Kangra and Hamirpur) were taken up for study (Plate. 21.1)

Remote sensed data has thus helped in rapidly generating natural resources information. It has also facilitated a continuous monitoring and recording of changes. With the use of computers and the GIS, it has become much easier to integrate various inter-related data including socio-economic data to arrive at optimum solutions and alternative strategies to meet the needs of the local people and maintain sustainability as best as possible.

However, it is not enough if such integrated data alone is made available and various action plans suggested. The local authorities, people and NGOs should actively pursue the implementation of the action plan, so that there is a sustained development activity that can be regularly monitored.

Funding for this is made available partly through the Department of Space, Government of India, various state governments and from the Union Ministry of Rural Development.

However, in addition to the availability of remote sensing and GIS technologies, local authentic information and co-operation from local officials and people is necessary to make the process successful. Whether sustainability will result in all cases is yet to be seen. With the availability of better quality and higher spatial resolution, satellite remote sensed data in the near future could lead to better and more detailed information systems to enable micro-level planning.

22

Preventing Natural Disasters

Remote sensing satellite information helps minimise losses in times of natural calamities by early warning systems.

Often, the words hazards and disasters are used synonymously. But hazards pose a threat to humans, while disasters cause destruction. Recognising a hazard and taking preventive measures can minimise losses.

Natural hazards range from the predictable to the unpredictable. Their causes could be one or many. Some are caused by geological and geomorphological factors, while others are due to environmental and climatic changes. The non-sustainable activities of humans has now become a major cause of natural disasters. Aerospace technology, particularly satellite remote sensing, helps in monitoring global changes and give early warnings of natural hazards. There are several international scientific bodies such as the International Committee of the Scientific Union (ICSU) which have proposed several scientific studies under the International Geosphere Biosphere Programme (IGBP). Satellite remote sensing is used to gather data in these studies.

Some typical hazards can be grouped under the following broad categories:

i) Exogenous—floods, drought, landslides, avalanches.

ii) Endogenous—eruption of volcanoes, earthquakes.

iii) Anthropogenic—subsidence (from activities such as underground mining), reservoir quakes, collapse of structures like dams, dikes, bridges and large buildings.

There are other hazards such as forest-fires, crop pests and diseases, locust invasion, tornadoes, etc.

Some disasters are very sudden, while some others are gradual and slow. For example, earthquakes, accidents, landslides are very sudden and last only for a few seconds or minutes. Some like forest fires may not be very sudden. Cyclones and floods, on the other hand, build up gradually. Disasters like drought, epidemics are noticeable only after a few months.

We can thus infer that some natural disasters can be prevented, while the effects of others can be mitigated by regular monitoring and timely action. Aerial photographs, radar images, satellite images of the Earth and information from weather or meteorological satellites like the Advanced Very High Resolution Radiometric (AVHRR) help in monitoring the physical changes on the Earth.

Many issues concerning our environment, such as the loss of biodiversity, erosion of forests and soils, pollution of ground water, climatic change due to the emission of carbon dioxide and other greenhouse gases are causing worldwide concern. According to the office of the United Nations Disaster Relief Coordination (UNDRO), approximately 90 percent of all natural disasters occur in the Third World countries. It is in this context that remote sensing technology should be considered a powerful and vital need for the Third World countries.

Establishment of an early warning system, selection of advanced technologies, dissemination of existing and new information on assessment, prediction, prevention and mitigation of such disasters will be crucial in the coming years and requires a multi-disciplinary approach.

While aerial remote sensed data was used for a long time, it is the satellite based remote sensing data which, because of its continuous availability and capacity to observe large areas, is considered a powerful medium to monitor the changes in the Earth's environment.

In India, the use of remotely sensed data for various hazard reduction studies has been highly encouraging. Aerial photographs have been used particularly in studying potential areas of landslides, both in the Himalayan region and in the Nilgiri hills in Tamil Nadu. Several soil conservation programmes were recommended and implemented successfully.

In India, floods occur mainly due to heavy rainfall in association with low pressure depressions and cyclones. While floods and cyclones cannot be totally eliminated, careful monitoring and planning can certainly mitigate the destruction and help in evolving suitable rehabilitation measures based on remote sensed data. While some state governments are trying to use this technology, there are many who have still not felt the need for use of remote sensing and flood-zone mapping as a long-term effort.

Certain types of disasters cannot be predicted directly through remote sensing techniques but potential zones can be identified fairly accurately. A careful study of fault zones, their types and disturbed landcover provide evidence for unstable zones. The boundaries of tectonic plates and laterally moving faults are areas of potential earthquakes. Recent studies suggest that microwave remote sensing data may provide some clues to potential earthquake zones, although there are no definite reports on the findings. Attempts are being made to evolve a drought prediction system using remotely sensed data, but drought prediction is difficult, and a foolproof mechanism will probably take time. However, the severity of droughts can easily be assessed, thus providing information to the authorities for implementing relief measures.

Forest fires occur frequently in tropical countries due to various reasons, particularly in the dry and hot seasons. Himachal Pradesh, Uttar Pradesh, Jammu and Kashmir are highly susceptible to forest fires. While some control and preventive measures are taken in some parts of our country, satellite remote sensed data is still to be recognised by the authorities as a powerful source of information. Forest fires are considered a potential hazard with physical, biological, ecological and environmental consequences. Since the number of forest fires are increasing every year, continuous monitoring is of great importance, not only to understand present trends but also to devise a model to predict the possibility of fires in future. The Rajiv Gandhi (Nagarhole) National Park situated in South Karnataka is well known for its biological diversity. In a recent fire, remote sensing satellite data was used and the results were very encouraging (Plate 22.2).

There is no system to forecast landslides, except that as a preventive measure large structures are not built on hill slopes to avoid damages. Soil conservation measures in areas susceptible to landslide could be identified and control measures undertaken.

Monitoring oil pollution is important because of the damage caused to marine resources by oil slicks. Satellite remote sensed data provide very valuable information in this area too (Plate 22.1 and 22.3).

In conclusion, remote sensing technology can help us by:

a) identifying disaster prone areas

b) predicting or forecasting, to some extent, disasters

c) monitoring natural phenomena to predict their onset and progress

d) assessing the damages and providing information for post-disaster management activities.

It is absolutely necessary to create adequate awareness among the public, as well as politicians and decision-makers on the benefits of remote sensed data.

23

Opportunities Open-Up

The IRS-1C, India's new generation remote sensing satellite, is expected to open up new possibilities in predicting natural hazards, hydel projects and even improving the prospects of the real-estate business.

With the new generation remote sensing satellite, IRS-1C, with improved sensors, application potentials could be enhanced. The fabrication of IRS-1C is complete and it has already been airlifted to the Baikanur Cosmodrome in Russia. Indications are that scientists and engineers have taken all precautions to ensure its successful launch.

IRS-1C is expected to help application scientists in planning and managing India's natural resources. In terms of spatial, spectral, temporal and radiometric resolutions, this will be superior to most satellites in the world, now. Besides these, it will have a higher resolution panchromatic camera which can be steered 26 degrees to look sideways. This will provide stereo images to measure elevations, which is at present done only by the French satellite SPOT. But SPOT's spatial resolution is only 10 m, whereas IRS-1C will give data at 5.8 m spatial resolution. The three sensors that the IRS-1C has are:

A Pan camera providing data at a spatial resolution of 5.8 m, in black and white picture form in a single spectral band of 0.5 to 0.75 microns (one micron is one millionth of a metre), covering a ground swath of 70 km. This camera will enable us to obtain a 5 day revisit (covering the same area) and also stereoscopic coverage. (To see the same area from two different view-points and to study the relief, as in the case of aerial photographs).

LISS-III, a medium resolution multi-spectral sensor providing data in four spectral bands, two in the visible (0.52–0.59 and 0.62–0.68 microns); one in near infrared (0.77–0.86 microns) and one in the shortwave infrared (SWIR-1.55 to 1.70 microns). It has a spatial resolution of 23 m (69 m in the case of SWIR band only) which will cover a ground swath of 141 km (148 km in the case of SWIR).

A widefield sensor (WIFS) provides data in two bands (visible 0.62–0.68 microns and near infrared 0.77–0.86 microns) but with a much larger ground swath of 810 km and at a spatial resolution of 188 m. These specifications were arrived at after elaborate and extensive discussions with application scientists with the country's diverse needs in mind (See Fig. 23.1).

Also, the satellite is truly global as it can observe any part of the world, record the data with an on-board tape recorder and download it to any ground station.

As of now, the Eosat Company of US, has a contract with the Department of Space, to receive and distribute IRS-1A and 1B data worldwide. The same arrangement will hold good for IRS-1C also. Certain other countries have also shown interest in direct reception of IRS-1C data through their own Earth receiving stations.

As for applications, data is needed to monitor once in a few days highly dynamic and rapidly changing phenomena such as vegetation dynamics, which at present has not been met adequately. Secondly, the pan camera will provide better spatial resolution data needed for mapping and monitoring urban areas and its growth. Generally a sensor with higher resolution will have a smaller field of view (ground swath) which needs a larger number of orbits for total coverage. Whereas, a sensor with lower resolution will have a larger swath and need a lesser number of orbits for global coverage. The payloads or a combination of sensors in the IRS-1C have been designed to meet these needs to a great extent.

While the feel of the data can be had only after it starts arriving, the application possibilities can be analysed with experience from past satellite remotely sensed data. To ensure continuity of data from IRS-1C, another satellite called IRS-1D will also be launched soon with similar sensors. The IRS-1C is expected to orbit at an altitude of 817 km, with the local time of crossing the equator being 10.30 hours.

A major application is in crop monitoring and yield forecasting. Besides crop stress monitoring and separability analysis using the high resolution pan camera, multi-spectral LISS III with improved spatial and spectral resolution and a wide field sensor, will become easier and more reliable. The pan camera data with a 5.8 m spatial resolution, will easily provide data for mapping on 1:25 000 scale and even larger, as at present we have not been able to map larger than the 1:50 000 scale. In addition, the camera will provide stereo pictures, which will help to perceive relief of terrains and facilitate the production of DEM. It will be possible to view the terrain in perspective, using appropriate computer software which is already available. The IRS-1C data will bring out terrain features better than the French SPOT is doing now. Smaller watersheds will be delineated better for detailed planning, essential for rural development and better water and land management, as the individual land holdings in India are much smaller (of the order of a few hectares).

The future lies in GIS and its extensive use in supporting various management decisions and aiding detailed planning to increase agricultural productivity. Data from IRS-1C will certainly go a long way in meeting these challenges.

DEM sometimes called Digital Terrain Model (DTM) will enable identification of areas potential for mini-hydroelectric projects in hilly regions. DEM, it is claimed, will also assist in studying environmentally hazardous regions, areas prone to natural hazards, and also to some extent indicate areas highly prone to earthquakes in conjunction with other supplementary information.

Another important application would be to create information for strategic planning for our defence forces particularly in the Himalayan regions. International boundaries and their demarcation can be done precisely.

Better soil resources inventory is expected, which in turn will improve the agricultural productivity according to soil scientists. In urban areas, the use of higher resolution data is bound to enhance planning processes with more detailed maps and more real-time information. Commercially, a variety of real estate businesses may grow faster. In addition, detailed flood mapping, landform analysis, identification of shoreline changes, will all improve substantially. In particular, LISS III sensor data will assist in the identification of vegetation types mixed crop situations and identify 'hotspot' bio-diversity indicators and wildlife habitat. This data is expected to be useful for better mineral exploration.

The WIFS data, with a broad coverage at about 185 m ground resolution is bound to be useful in understanding vegetation dynamics, drought monitoring (supported by other supplementary information), sediment movements and monitoring forest fires.

IRS-1C data will be a quantum jump in providing information on India's resources in near real-time. It is expected to open up more challenges in managing the country's natural resources and monitoring the changes in landuse practices. It will help in taking corrective measures faster than what has been possible so far. It may also be noted that in the near future, 1 and 2 metre resolution satellite data is likely to become available from western countries and with such data in combination with IRS-1 data, may lead to the development of a much better and precise information system which will become a need for improved living conditions in developing countries. In fact, the total system of measuring, mapping, monitoring and managing Earth resources and planning development will slowly shift to satellite-based geotechnology with reduced ground visits and use of aerial photographs which is more expensive.

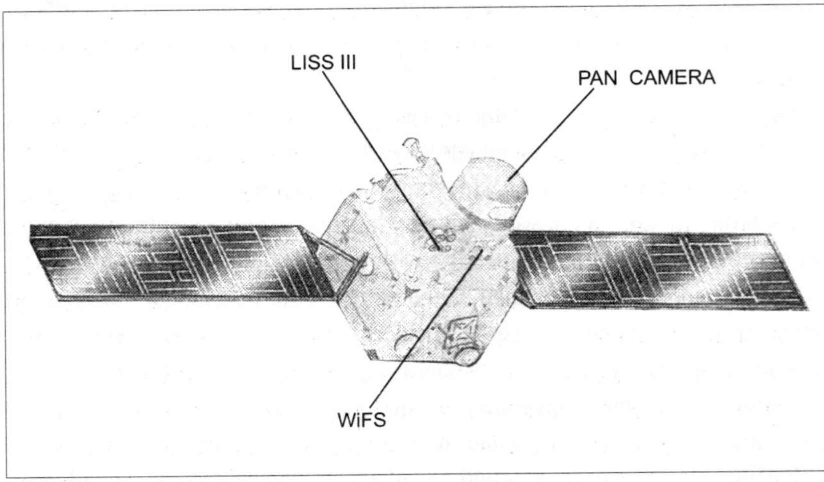

Fig. 23.1 *An artist's concept of IRS-1C in orbit*

24

Careers in Remote Sensing

This brief examines how space technology is taught at Universities in India and the necessary opportunities to make a career in Remote Sensing.

In any new technological endeavour and development, the application aspects always take time to catch up. This is particularly so in the case of remote sensing technology and its applications. Space technology is fast advancing both in terms of quality of data as well as improvements in spatial, spectral, temporal and radiometric resolutions. Data with, 2m or even 1m measurable resolution will appear before the turn of the century. The people who make use of these data will have to be aware of the various implications, and decide on the fastest and most economic route they should adopt to achieve their goal.

This article attempts to highlight educational and training opportunities available in India and the extent to which they can serve the purpose effectively, besides mentioning the working facilities available.

Till recently, most academic activities were confined to postgraduate diploma or degree level courses. Remote sensing was essentially used to survey land for both its dimensions and for the resources it bears. Therefore, a person who wishes to become conversant with remote sensing and its applications will need a sound knowledge and experience in his own field of specialisation, such as basic land survey and measurement knowledge. He must know about aerial photography and other data collected, geological mapping, soil mapping, water resources mapping, forestry survey and mapping, land use and land cover mapping. The data collected by remote sensing medium can then be used to the best advantage and to obtain quick and optimum results.

Surveying was taught in the undergraduate programme to a limited extent as part of the civil engineering curriculum, but the importance given to it was meagre. With the development in the use of aerial photographs, not only for making measurements but also for resource studies, the need for trained persons was felt, but as the aerial photographs were classified and not easily accessible, their use was confined to those who had the access and authority to use them.

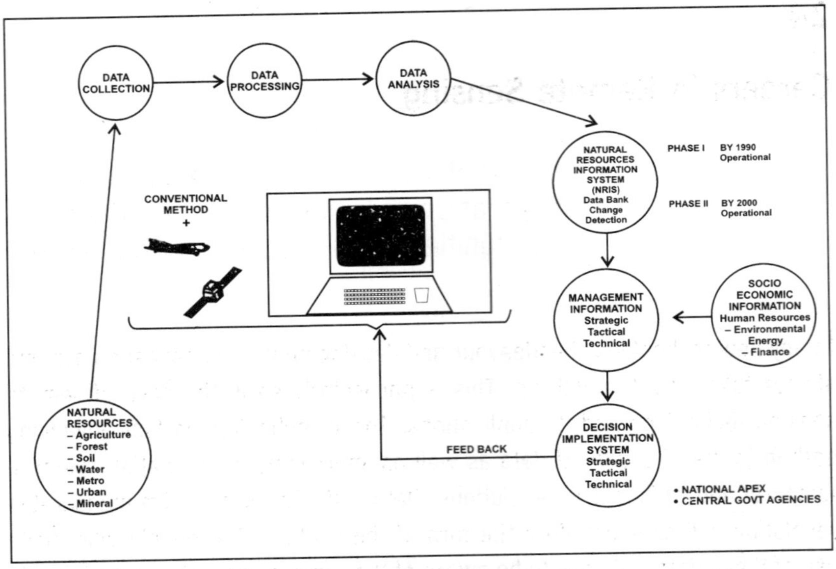

Fig. 24.1 *National Natural Resources Management System. (Source: ISRO)*

The first institution for training in-service professionals or resource scientists in the use of aerial photographs began in Dehra Dun as a part of the Survey of India in 1966 and was named the Indian Photo Interpretation Institute, a project established with the active support and collaboration the International Institute for Aerospace Surveys and Earth Sciences of the Netherlands. Besides this, the Survey of India had its own institution for training various categories of people for its own surveying and mapping work. The Indian Photo Interpretation Institute in 1976 or so came under the National Remote Sensing Agency and was renamed the Indian Institute of Remote Sensing. It was the main training wing and still is one of the premier institutions for training in the use of application of remotely sensed data. It admits scientists, teachers from universities, officers deputed from the Forest and the Soil Survey departments, water resources engineers and land use planners. There are different levels and durations of training in the use of this modern technology. Trainees from foreign countries too are admitted.

Private candidates perhaps may also be entertained but the fees may not be affordable for many. The main aim is to train resource scientists in all aspects of remote sensing, so that when they go back to their parent department, they can apply this technology in their day-to-day work. This institute cannot be called large in numerical terms, as its facilities are limited.

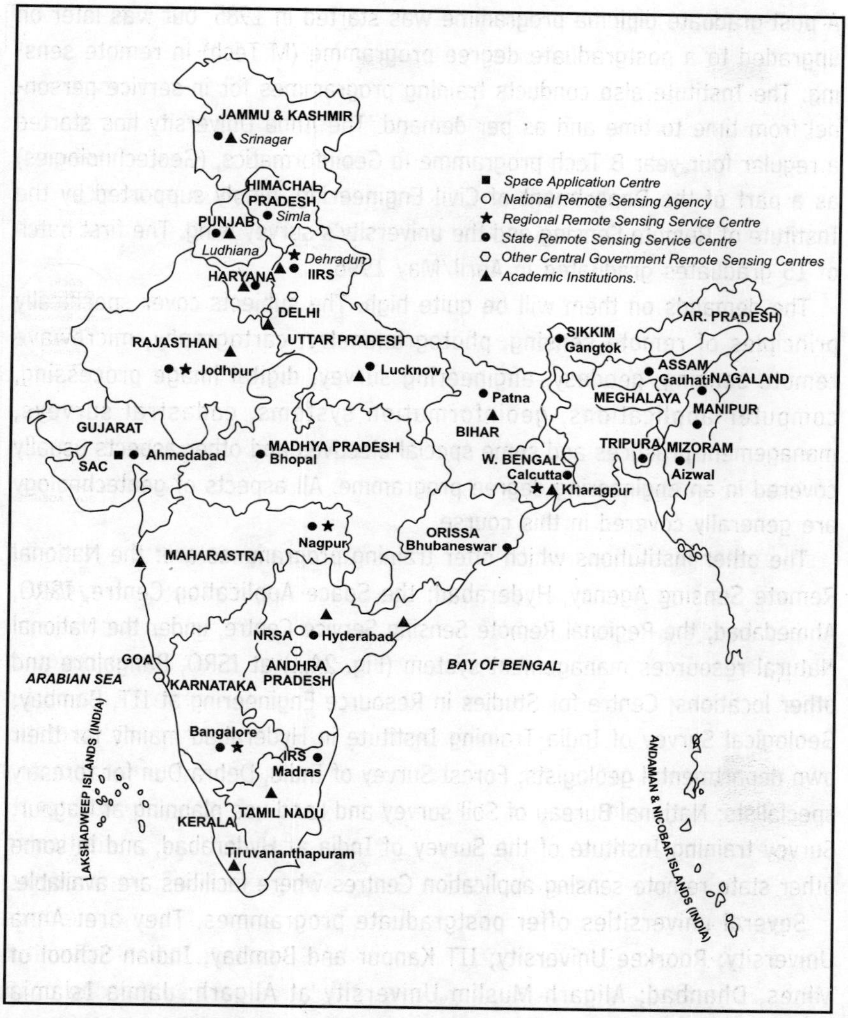

Fig. 24.2 *Remote sensing facilities in India. (Source: ISRO)*

Recently it was announced that the United Nations, under one of its programmes of supporting two or three centres of excellence in different parts of the world in this subject, has chosen the Dehra Dun Institute to cover this part of the world.

Another important centre where both educational and training aspects are covered is the Anna University, Madras; the Institute of Remote Sensing was started here in 1982. It plays the dual role of rendering service in the form of paid-for remote sensing application projects by those who make a request, and also supports the university's regular educational programmes.

A post-graduate diploma programme was started in 1985, but was later on upgraded to a postgraduate degree programme (M Tech) in remote sensing. The Institute also conducts training programmes for in-service personnel from time to time and as per demand. The Anna University has started a regular four year B Tech programme in Geoinformatics, (Geotechnologies) as a part of the Department of Civil Engineering, amply supported by the Institute of Remote Sensing and the university's survey wing. The first batch of 15 graduates graduated in April/May 1996.

The demands on them will be quite high. The subjects cover specifically principles of remote sensing, photogrammetry, cartography, microwave remote sensing, geodesy, engineering survey, digital image processing, computer applications, geoinformation systems, cadastral surveys, management practices and some special electives and other aspects usually covered in an engineering degree programme. All aspects of geotechnology are generally covered in this course.

The other institutions which offer training programmes are: the National Remote Sensing Agency, Hyderabad; the Space Application Centre, ISRO, Ahmedabad; the Regional Remote Sensing Service Centre, under the National Natural resources management system (Fig. 24.1) at ISRO, Bangalore and other locations; Centre for Studies in Resource Engineering at IIT, Bombay; Geological Survey of India Training Institute in Hyderabad mainly for their own departmental geologists; Forest Survey of India, Dehra Dun for forestry specialists; National Bureau of Soil survey and Land use planning at Nagpur; Survey training Institute of the Survey of India at Hyderabad, and in some other state remote sensing application Centres where facilities are available.

Several universities offer postgraduate programmes. They are: Anna University; Roorkee University; IIT Kanpur and Bombay; Indian School of Mines, Dhanbad; Aligarh Muslim University at Aligarh; Jamia Islamia University at New Delhi; B M Birla Science and Technology Centre at Jaipur; Andhra University, Visakhapatnam; Bharathidasan University, Tiruchi; Regional Engineering College, Warangal; Jawaharlal Nehru Technological University, Hyderabad and the Madras University. Many universities offer remote sensing as a part of the course in subjects like geology, geography, civil engineering, hydrology, geophysics, forestry, architecture and computer sciences.

However, one drawback in a number of these institutions and universities is that the syllabi or details taught are not uniform, in the case of regular post- graduate degree programmes, and the students are also not uniformly educated and informed. In other words, there is no co-ordination.

It is hoped that the situation will improve and well informed candidates will emerge, with improved standards. The UGC and the Department of Space should take more initiative in this respect, although a curriculum seems to have been drawn up.

At the school level, there is a small element of remote sensing included in the higher secondary level. But it has not been effective as there are not many informed teachers, and no proper textbooks. The Department of Space has taken some initiative in this regard, but much more needs to be done.

Almost all States have started remote sensing application centres; while some are autonomous, some are part of other departments. Most of the state centres are not well funded and have only limited staff. However, the Department of Space has set up five or six regional remote sensing service centres, all of which are well equipped and have major facilities like NRSA, Space Application Centre and ISRO. There are some Central Government organisations such as the Central Water Commission, Oil and Natural Gas Commission, Forest Survey of India, Central Arid Zone Research Institute, Atomic Minerals Division, Geological Survey of India, etc. with their own facilities.

While adequate facilities have been built up, it is questionable whether all these are being optimally used and whether adequate manpower and support is available or not. If such high technology and high cost facilities are created but not fully used and managed, it would be a pity. Many Government departments need information generated by remote sensing and GIS, but most of them do not want to support institutions which have already been established. It would be desirable that when budgetary allocation to various user departments are made, a small proportion of five to ten percent should be earmarked for supporting remote sensing facilities. This concept is easier said than done, but it deserves some consideration.

The map shows the spread of various facilities in India, duly marked with all details as extracted from a recent Department of Space publication (Fig. 24.2).

25

Remote Sensing for the Military

Satellite information can be used for military as well as civilian purposes. The Indian authorities must quickly update their systems to keep up with the latest in the world.

Our President, while presenting colours to the Flying Instructors School of the Indian Air Force at Jodhpur on October 26, 1995, emphasised the need for greater military preparedness. He said the sub-continent was threatened by terrorism and extremism. There could be no compromise in this regard and there should be no complacency in dealing with matters of national security. Collecting information (or intelligence) and keeping it constantly updated is a very important component of being prepared. The role of remote sensing in military preparedness will be examined in this article, keeping in mind the present state of technology and satellite systems available the world over.

In an article published recently in the official journal of the *Indian Society of Remote Sensing* (Vol. 23, No. 3, Sept. 1995) B L Deekshatulu, Director NRSA and M G Chandrasekharan of ISRO, headquarters, have traced the emergence of remote sensing technology as a system for obtaining information, or for spying by Russia and the US particularly, to observe each other's military installations and for monitoring missile deployment.

The first US effort with the launch of the Discovery Series in 1959, followed the by Area Surveillance Systems (SAMOS) have been mentioned. The Soviet Union in turn launched the Kosmos 4 in 1962. Subsequently, several countries tried to enter into similar activities. In the course of time, the transformation from military to civilian uses came up and today there are several satellites in orbit and several more are planned throughout the world.

The recent state-of-the-art and most versatile civilian satellite is the Indian Remote Sensing Satellite (IRS-1C), which was successfully launched on December 28, 1995 with a spatial resolution of 6 m in black and white and multi-spectral linear imaging self-scanner data with a 24 m spatial resolution (Fig. 25.1). With the facility to tilt the panchromatic (black and white) camera on either side, we are now in a position to obtain data every fifth day with the capacity to view them stereoscopically to observe relief or heights in terrain conditions. (Plate 25.1)

How the Indian military is getting ready to use remote sensing, or is already doing so, is not clearly known as this is classified information. But it would be absolutely necessary to use it for gathering military intelligence as well as keeping it up-to-date on a near real-time basis, both for strategic planning as well as tactical planning to some extent.

In an item published in the *Space News* (Vol. 6, No. 21) (May-June, 1995) in the US, a senior Indian defence official (name not given) was quoted on how India is using its space system for military purposes. This paper says that India's space and defence research establishments have begun a new military space project that will bolster India's electronic surveillance capability. It also states that ISRO and the Dehra Dun-based Defence Electronics Research Laboratory (DEAL) is working on a range of satellite-based surveillance devices intended to shift the present ground-based intelligence gathering effort to a satellite-based one within the next five years.

It was also stated that ISRO would provide remote sensing satellites in low Earth orbit with "high resolution cameras to physically spy on countries from

which India sees a potential threat." It implies that senior defence commanders will be able to get nearly instantaneous battlefield information on troop movements and other military activities of rival nations. There is also a possibility of military satellites being launched after some years although for the present, data from existing satellites will be used.

The capabilities, in terms of spectral, spatial, temporal and radiometric resolutions of IRS-1C which was launched on December 28, 1995; how and in what manner the data obtained from this satellite is going to meet the actual needs of our military cannot be stated at present. The IRS-1C's Pan camera (black and white pictures) can get data with stereoviewing capability and with suitable software. It can easily provide a digital terrain model, or digital elevation model using the appropriate computer system, which in turn can facilitate perspective views of the terrain, particularly along the Himalayan region, with complete actual terrain features (Fig. 25.1).

This will also enable the development of a GIS, which can greatly enhance the military intelligence base, for strategic planning, combined with information obtained or available from other conventional ground-based or aircraft-based systems.

Table 25.1 Some Military Satellite Sensors

Sensor	*Functions*	*Type of Satellite*
Photographic	Detection of * ICBM – deployment * Cruise missiles deployment * Troop movements	US Optical reconnaissance satellites (Big bird, KH – 11)
IR	Detection of * Missiles * Aircraft * Cruise missiles (potential)	US and Soviet early warning satellites
Nuclear radiation detection	Detection of * Nuclear explosions in the atmosphere * Nuclear explosions in outer space	US Vela satellites and IONDS – systems on GPS – satellites
UV	Detection of * Fluorescence from boosters * Fluorescence from nose aeons (during ballistic flight)	US and Soviet early warning satellites
Electronic signal detectors	* Interception of radio and microwave transmission * Radar-signal detection * Missile telemetry interruption	Soviet COSMOS and US Elint satellite US Rhyolite and Chalet satellites (INGSO)
Radar	Detection of * Naval surface ships * Ground-based military objects	US Seasat 1 (1978) and Soviet COSMOS ocean surveillance satellites
IR (thermal) Scanners or radiometer	Night-time reconnaissance Detection of * Buried structures * Underground construction	US – Seasat 1

In a recent week-long workshop of one commission of the International Society for Photogrammetry and Remote Sensing (ISPRS) held at the Anna University from November 27 to December 1, 1995, Lawrence W Fritz of the Lockheed Martin Corporation of the US and who is also the Secretary General of the ISPRS gave an elaborate account of the activities in the US by nine private companies which are trying to launch private satellites on a commercial basis with resolutions of 1–3 m on the ground, and he also mentioned other countries like France, Germany, Israel and Japan and gave details of the resolutions that will be achieved in the coming years. The Early Bird commercial satellite and the Earth Watch World View are expected to deliver

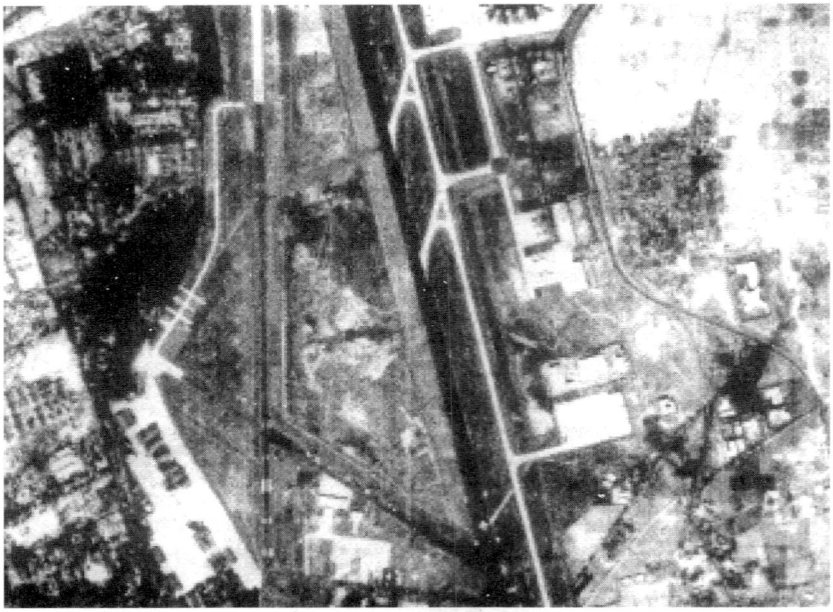

Fig. 25.1 *A satellite picture of Delhi airport*

data of 3 m resolution in 1996 along with 15 m resolution multi-spectral scanner data. In fact, photography and remote sensing have become one of the key elements in the spatial information sciences.

Since the break-up of the USSR a few years ago, the Russian satellite data with a spatial resolution of two or three metres became available commercially and the US government has now licensed large private companies to launch their own satellites and make very high resolution data commercially available. It appears that the market for such data is enormous, both for military and civilian uses.

One can, therefore, conclude that the Indian defence authorities should not lose any more time to develop their satellite remote sensing capabilities and enhance their intelligence capabilities more on a near real-time basis. The ISRO can develop satellites specifically suitable for military uses and also make data available more frequently, but how long it will take is anybody's guess. In addition to what is stated above, the defence authorities will perhaps need satellite data acquired in the microwave region of the electrogenetic spectrum to keep a closer watch of military developments around our country. It is hoped that our military authorities are quite aware of the development and will take to this technology quickly.

26

Remote Sensing for Managing Water Resources

Water is a precious commodity which is rapidly becoming scarce. This brief discusses the various water sources and how water management must be optimised and recharged to meet our growing needs.

Water is one of the essential natural resources without which we cannot exist. The demands for water made by agriculture, industry and for human and cattle consumption is increasing rapidly, particularly with the global increase in population. Therefore, the optimum use of water resources is absolutely essential and this can be done by adopting modern management practices and preventing waste caused by uncontrolled run-off. Shortage of water during periodic drought conditions in certain parts of the country has been tackled, but there is still a long way to go. There is a need to initiate studies and action to prevent loss by seepage, evaporation and run-off into the ocean when rainfall is excessive and rivers get flooded. We cannot still say that all the precipitation that occurs and all the water that several of our river systems carry during the monsoon is effectively tapped and stored both over and underground so that during the lean period water will still be available for a variety of uses.

Experts have suggested several measures that could be taken, but the action programmes drawn up have not been fully implemented for a variety of reasons. The Central Water Commission has published a booklet on water resources in India in 1987 but crucial decisions are yet to be taken as several constraints are foreseen today. According to this report, after allowing for evaporation losses, etc., the country's estimated water potential from surface flow is 1 800 cubic km and all this is carried by the various river systems in the four or five months in a year of the monsoon season.

Being a large country with a total land area of 329 million ha and with a population of nearly 900 million, unless the government takes serious measures and evolves a national sustainable water policy, there will soon be a severe shortage of water even for essential uses. All the water resources available, particularly surface water resources, should be considered as a national asset.

Groundwater is one of the essential sources in rural India. Using remote sensing techniques, potential groundwater sources have been identified and exploited. However, using power-driven pumps the extraction is much more than the quantum of replenishment or recharge, particularly in areas where rainfall becomes inadequate and drought conditions persist. While several measures are being taken, again by using remote sensing technology and constructing certain types of structures, no regular water balance study has been done to ensure that the recharge is nearly the same as extraction, so that the water level does not go down rapidly.

There has been no control worth its name in this aspect. As Dr P R Pisharoti, a meteorologist said, "Many people think that groundwater is an inexhaustible source. It is forgotten that most of the groundwater, accessible through deep borewells, is also a product of the rainfall in the past hundred years. The rainwater that does not run off the ground surface and the amount that does not evaporate penetrates into the soil to form groundwater. The recharging water sinks into the soil at an approximate of one metre a year. Therefore, the water that is available at 200 or 60 m below the ground surface is water that fell as rain about sixty years ago."

We have exhausted all that has fallen within that period. Replenishment of groundwater is poor even in good monsoon years. Therefore, it is imperative that groundwater is recharged at the proper places which have to be identified using remotely sensed data. This is much easier than conventional methods, particularly when there is excess rainfall and periodic flood conditions. There is also an urgent need to dig ponds as deep as possible, and divert rain water into them. Such activities apply even to the urban areas where water falling on roofs of houses could be diverted to locally devised ponds or low-level tanks suitably designed so that most of the water is not allowed to run-off, because our rainfall occurs only in a few months of the year.

A publication of UNESCO brought out in 1986 tells of how satellite remote sensing data could be a valuable tool to locate possible areas of artificial recharge taking a larger view of the world over, in addition to monitoring groundwater levels. Now that there are better quality satellite remote sensed data, more effort should be made to locate areas potential for recharge, potential for creating deep wells (inverted wells) and evolve continuous process of recharge efforts, so that we could have the most sustainable water usage possible.

Agriculture and irrigation are two areas which require water. Since Independence, this aspect has been well recognised and several large-scale

Plate 15.2 SPOT data of Madras urban sprawl.

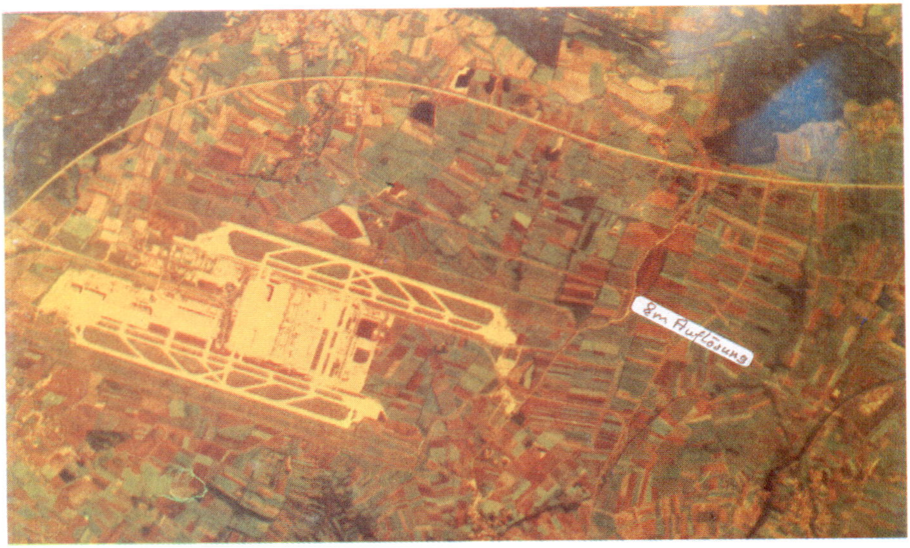

Plate 16.1 An airfield imaged by a Russian Air Satellite.

Plate 16.2 Chernobyl disaster in Russia.

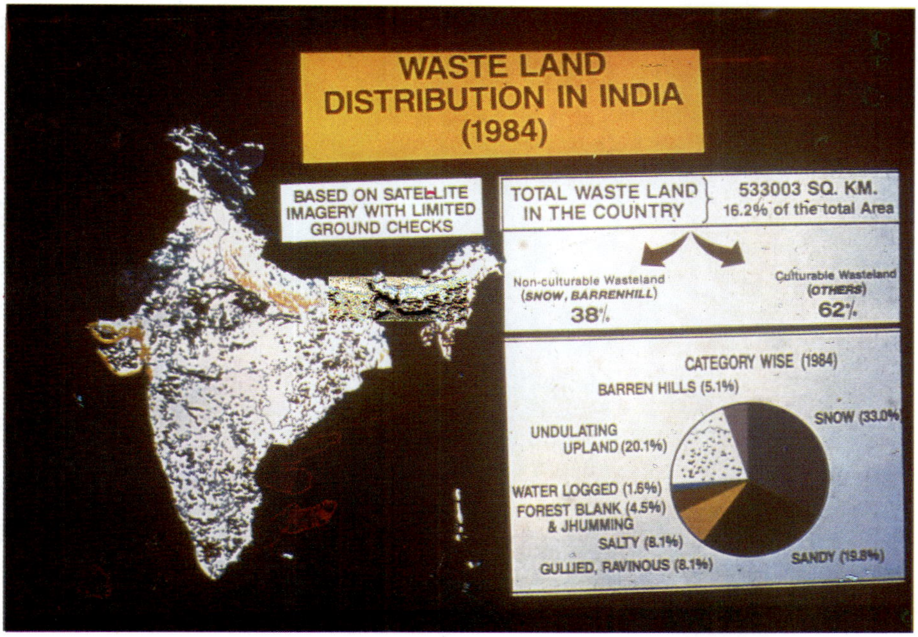

(a)

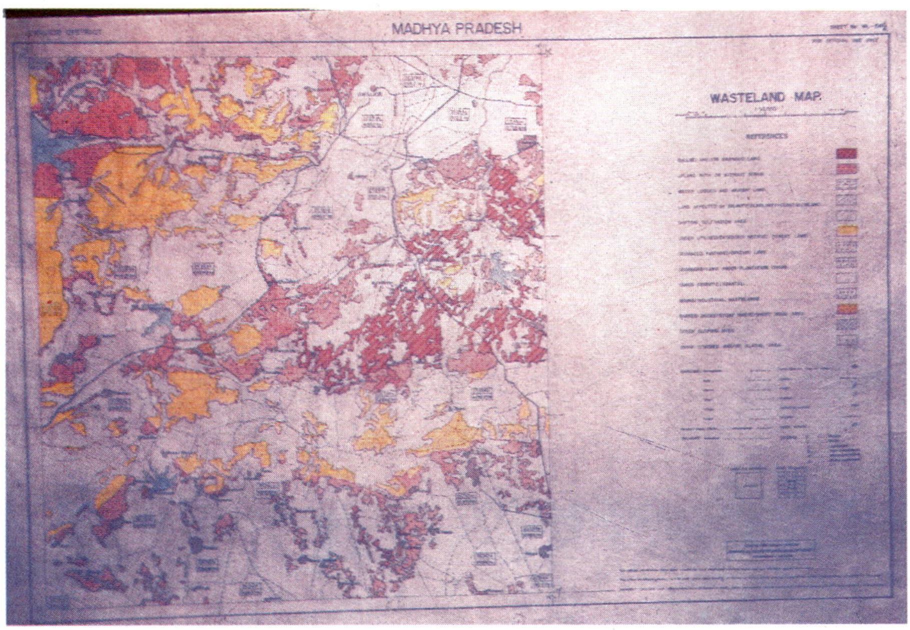

(b)

Plate 17.1 (a) and (b) Mapping wastelands through remote sensing. (Courtesy NRSA and ISRO Headquarters)

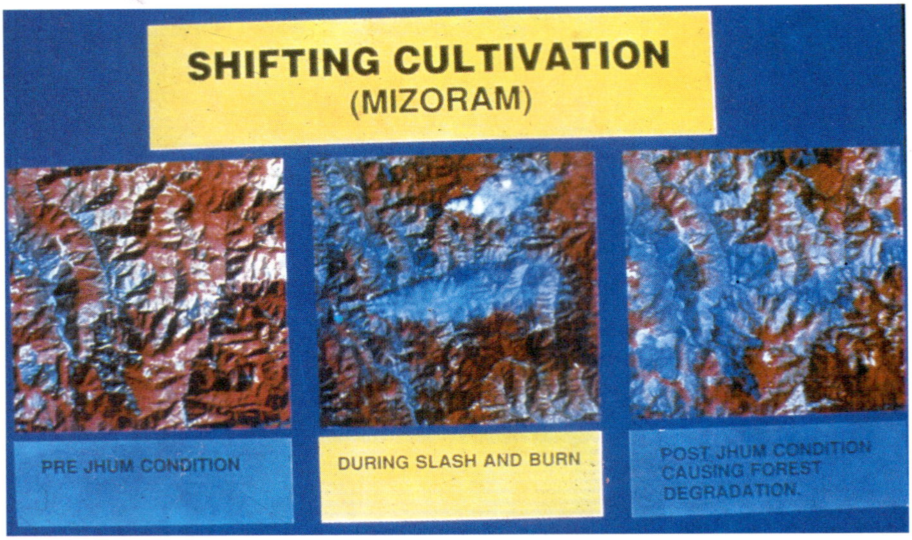

Plate 18.1 Shifting Cultivation, Mizoram.

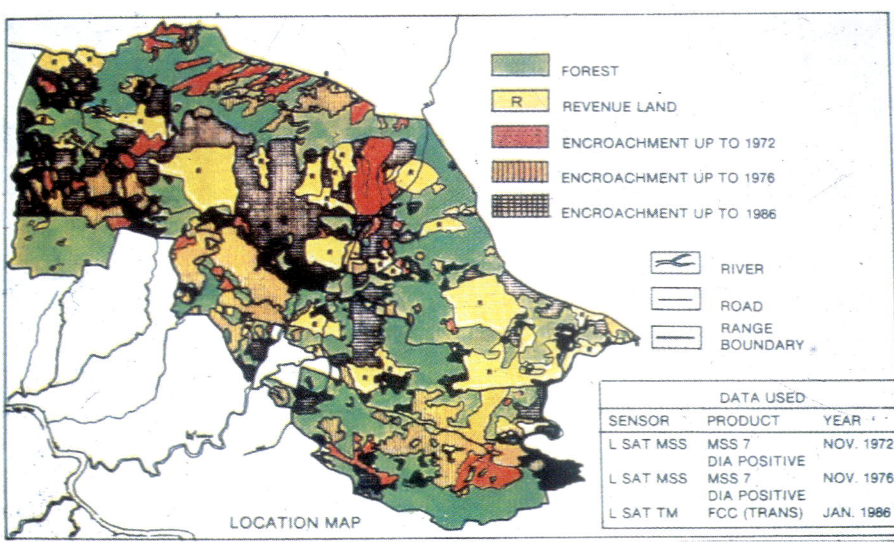

Plate 18.2 Encroachment spread, Shirpur and Sangvi ranges, North Dhule, Maharashtra.

Plate 19.1 The Indian space effort, from design to launch and information-gathering, is on par with the best in the world.

Plate 19.2 The IRS-1B: Indian remote sensing data may be the most coveted in the world by the year 2000.

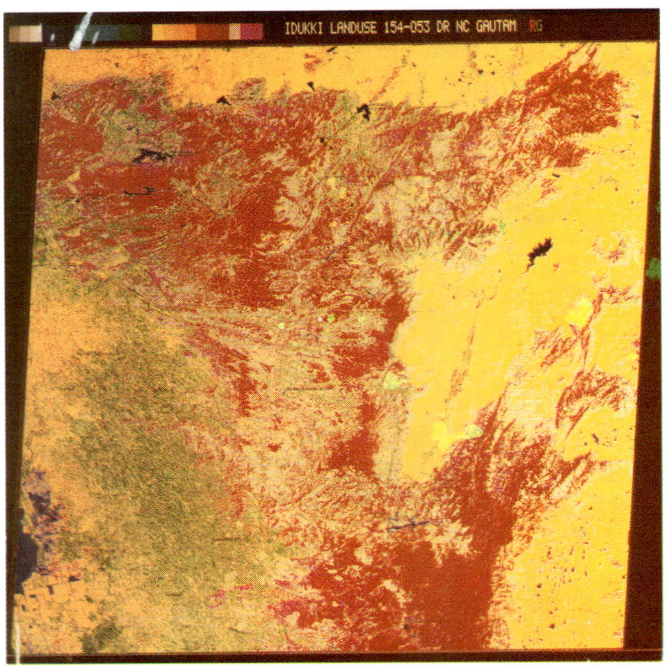

Plate 20.1 Land use configuration of the Periyar river basin in 1973, before the dam was built.

Plate 20.2 Land use configuration of the same river in 1981, after the dam was constructed.

Plate 21.1 False colour composite of a part of Chamba district.

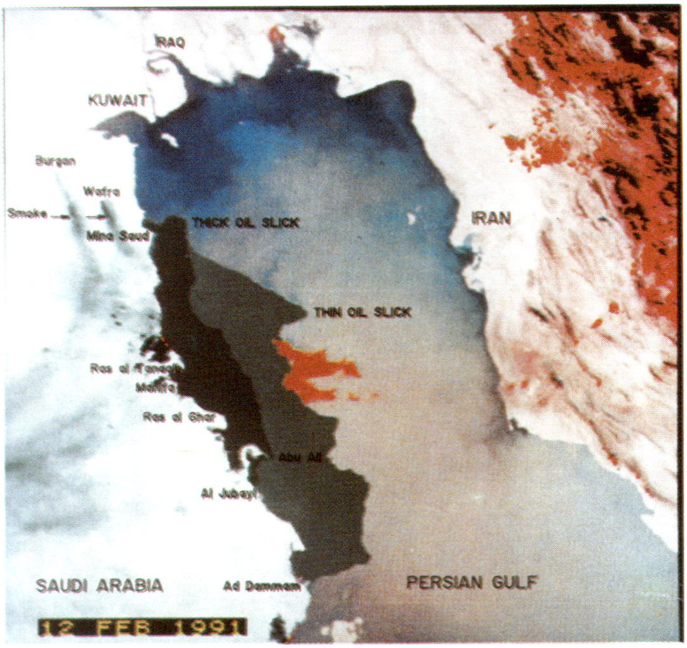

Plate 22.1 Remote sensing satellite picture dated February 12, 1991, showing the spread of an oil slick in the Persian Gulf.

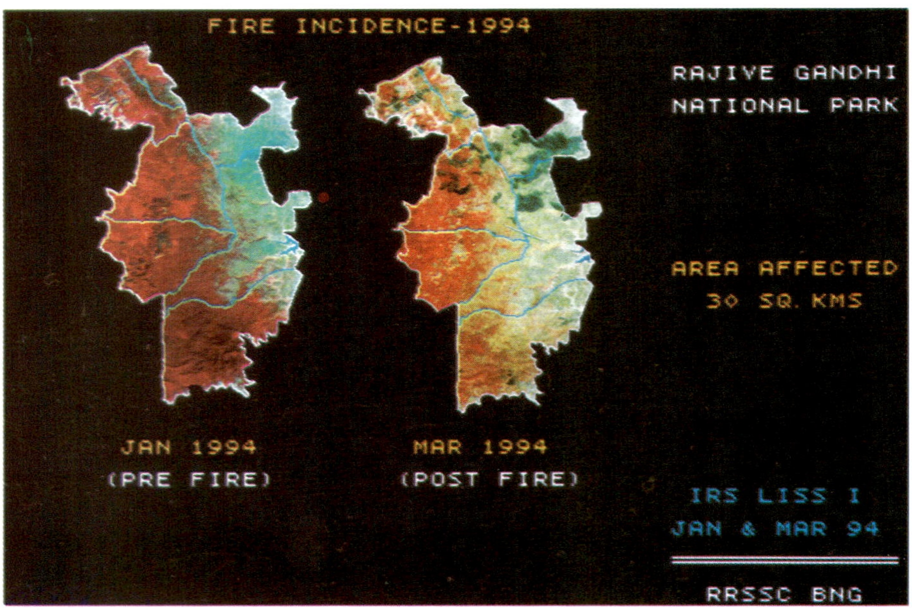

Plate 22.2 Satellite picture of a forest fire in the Rajiv Gandhi National Park which affected an area of 30 sq. km.

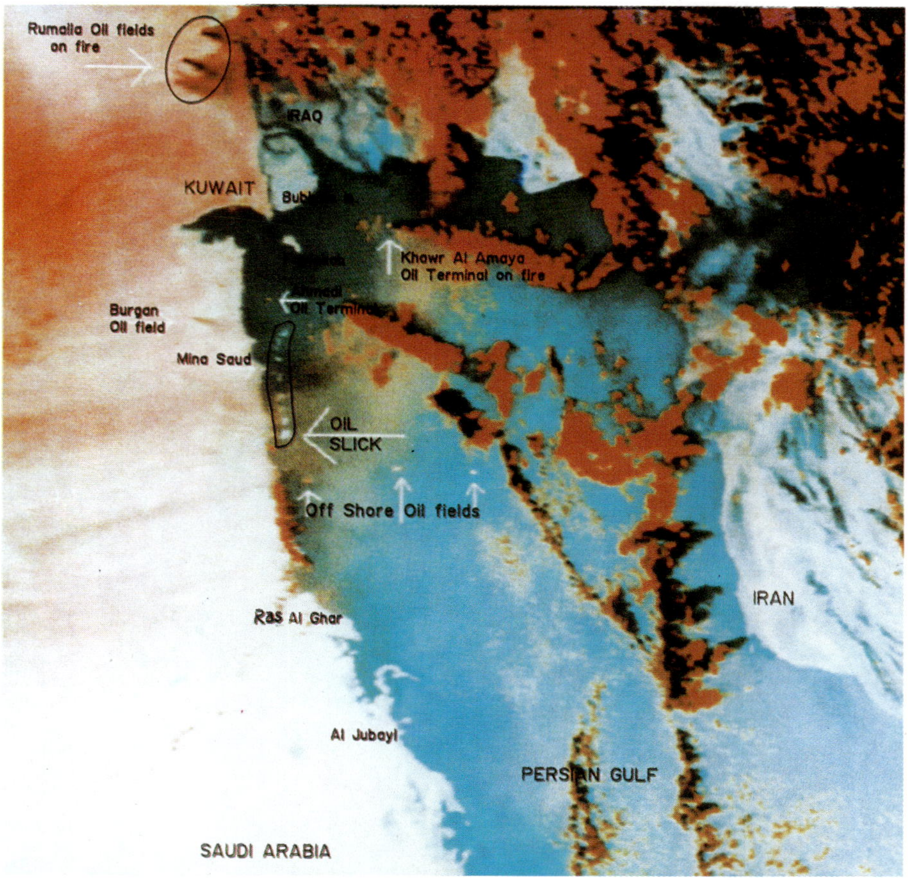

Plate 22.3 Remote sensing satellite picture indicating oil slicks and fire in the oil fields around the Persian Gulf.

Plate 25.1 First multi-spectral image from IRS-1C showing Goa and its environs.

Plate 26.1 Satellite image of 12 March 1994 overlaid with cadastral details in Bhadra Command Area.

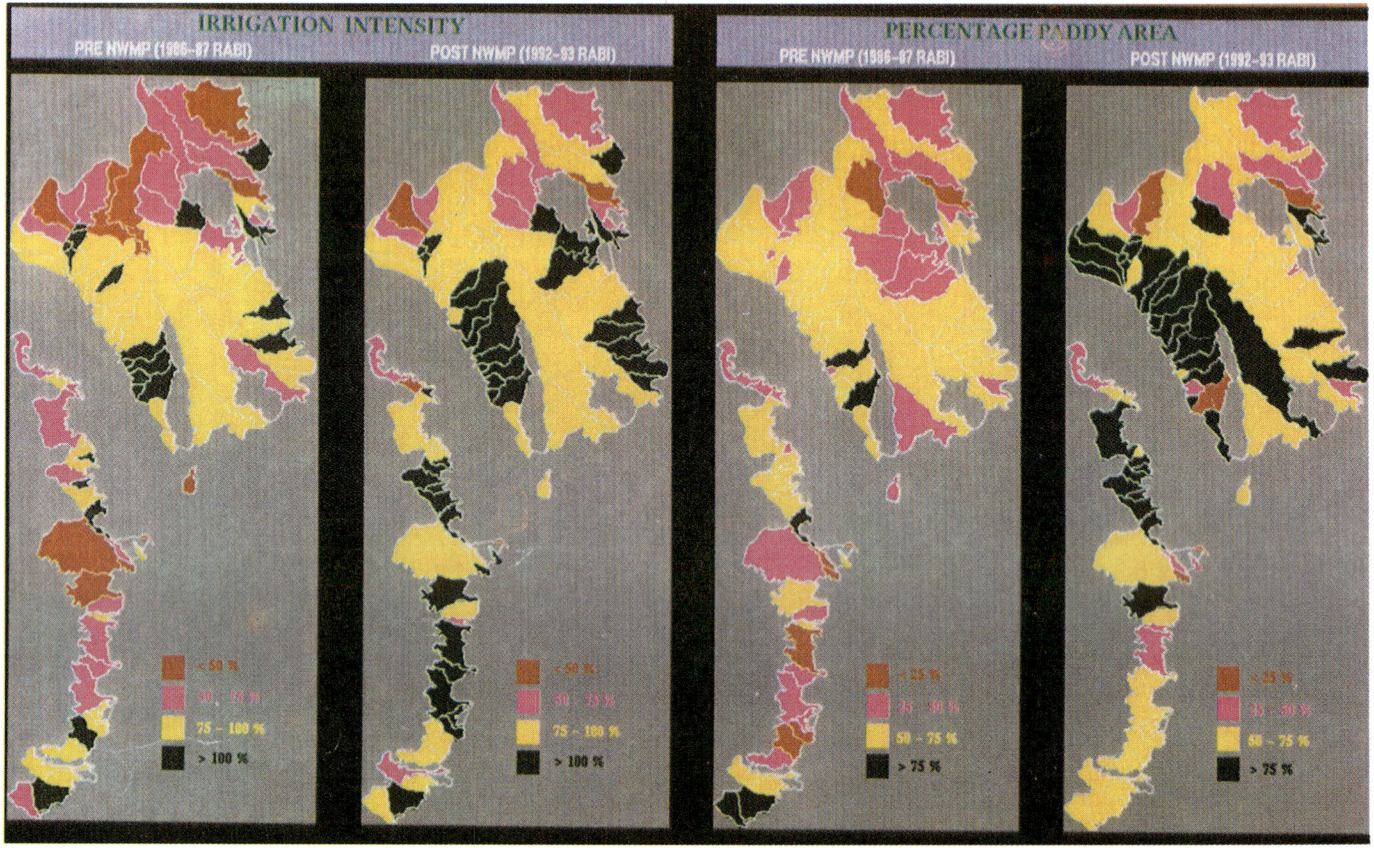

Plate 26.2(a) The Bhadra project in Karnataka: irigation intensity and percentage paddy area.

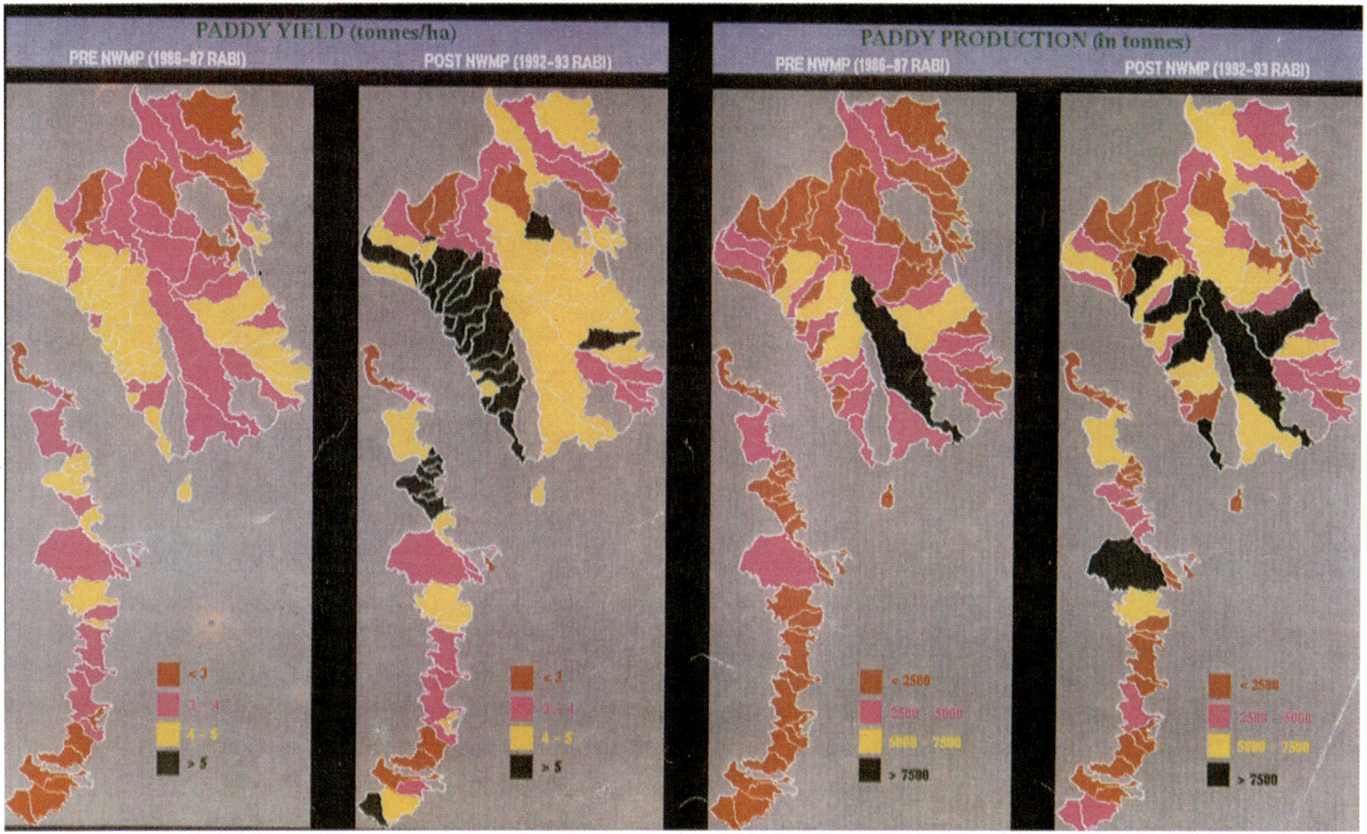

Plate 26.2(b) The Bhadra project in Karnataka: paddy yield and production

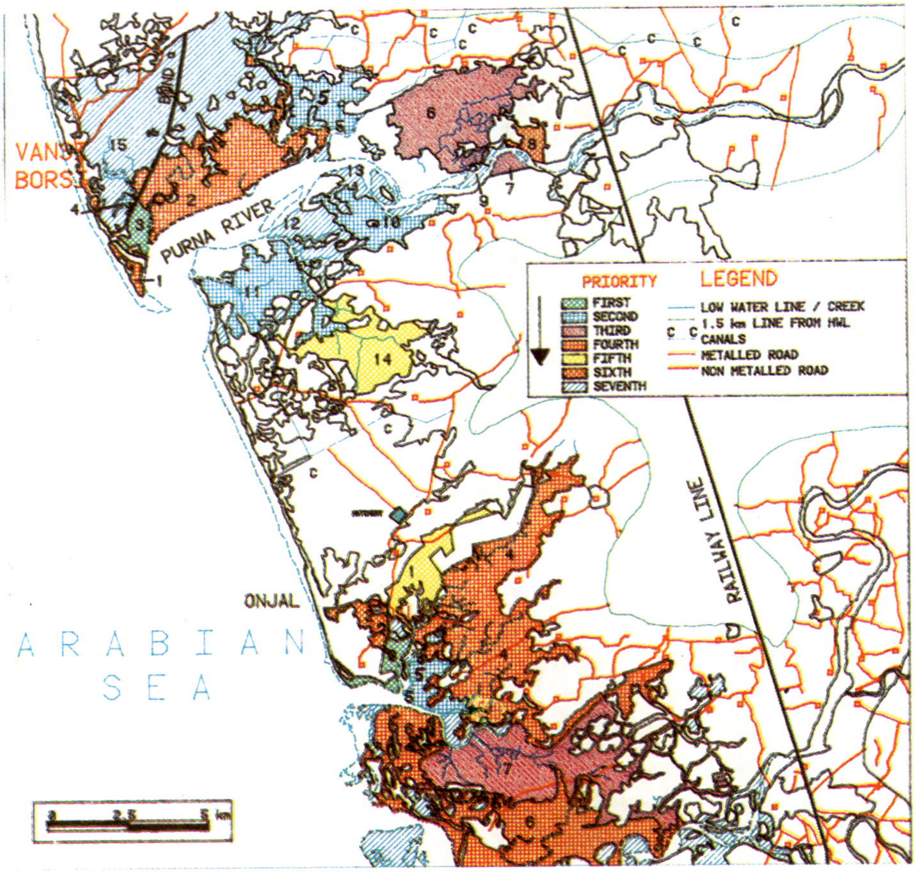

Plate 27.1 Sites for aquaculture in South Gujarat. (Courtesy SAC-ISRO)

Plate 28.1 Satellite picture of land formation in South India and Sri Lanka.

Plate 28.2 A remote sensing satellite picture of Goa and its environs.

Plate 28.3 A view of the land north and north east of Hyderabad. (Courtesy NRSA)

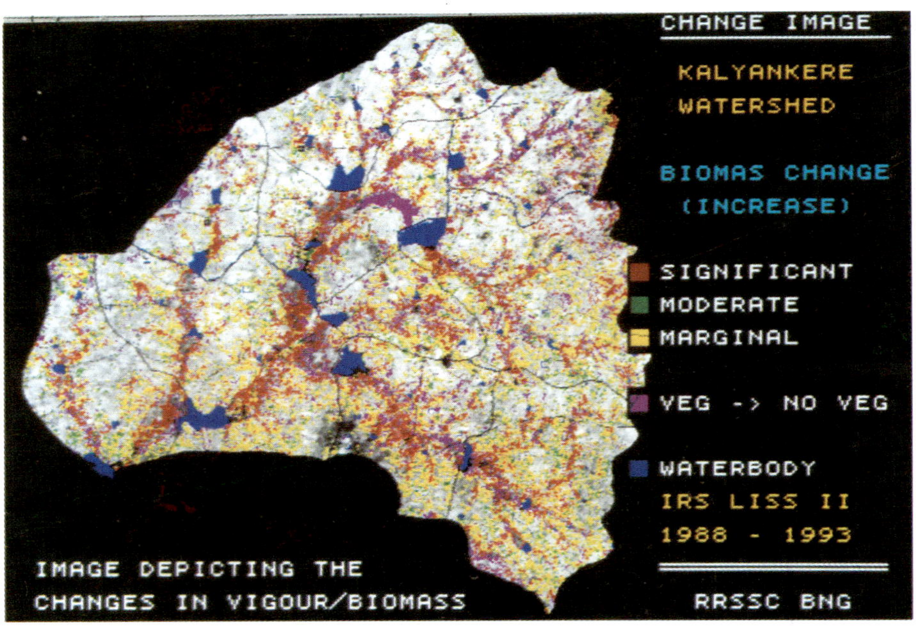

Plate 29.1 Image depicting the changes in vigour/biomass.

dams have been constructed anticipating certain benefits coming out of it economically, although it has not been as expected according to several reports. Whether we should go in for large multi-purpose dams or several small dams has been debated on several occasions. Again, conjunctive use of surface and groundwater for irrigation purposes has also been debated so that we are able to make full use of the water that becomes available through precipitation which varies in quantity as well as in locations.

In this context, we shall now look into a case study sponsored by the National Water Management Project under the overall irrigation management in canal commercial areas carried out by the water resources group of the National Remote Sensing Agency in 1994–95. The objective was to illustrate the possibility of increasing agricultural productivity and farm income through reliable, predictable and equitable irrigation service, and demonstrate the satellite remote sensing data capability to generate primary data on irrigated area, cropping pattern and crop condition and yield under distributory and sub-distributory levels.

This data along with relevant ground data have been analysed by NRSA scientists, to evaluate the improvement in system performance after implementation, and to diagnose problem distributaries of low irrigation intensity, deviation from national cropping pattern and low and stagnant yield and water use efficiency for corrective management. With the new IRS-1C satellite data, concurrent monitoring in near real-time through the season will also be possible with these new data systems that have now become available. The aim is to increase agricultural productivity and farm income in the existing water potential schemes through data that is reliable, predictable and accessible.

As of now, the satellite remote sensing techniques are being used to generate reliable and objective primary data on existing irrigation schemes and the resultant productivity. Ground surveys also have to be conducted and supplementary information obtainable from satellite data. Till now data obtainable from the Landsat series of multi-spectral data and thematic mapper data with a spatial resolution of 80 m and 30 m respectively at a 16 day repetitive coverage were used. Of late, since the launch of the Indian Remote Sensing Satellite, data with 72.5 m and 36.5 m resolution was used from 1988 to 1993. These provided a 11 day repetitive cycle of data available, except when there is cloud cover.

Now that the IRS-1C was launched on December 28, 1995, and is ready to supply data to users it is expected to be used, as its multi-spectral sensor is providing data at a spatial resolution of 23 m with a repetitive cycle of 24 days, with a swath of about 140 km. The panchromatic black and white data of 6 m ground resolution with stereo viewing capability and also the possibility of 5 days repetitive cycle will be of great use, with a ground swath of 70 km. There are other Indian remote sensing satellites of the 'p' series launched by our own Polar Satellite Launch Vehicle, beaming data which are useful for such studies.

In addition, the European Remote Sensing Satellite (ERS-2) gives data using a synthetic aperture radar with cloud penetration capability and a spatial resolution of around 30 m. The use of such multi-satellite remote sensed data will be an advantage for several uses including water resources studies in its several facets.

In the case of the Bhadra Project (See Plates 26.1 and 26.2 (a), (b)) in Karnataka, the irrigation intensity in this recent rabi season was found to have increased to 90 percent compared to 76 percent before the use of the National Water Management Project. Similarly, in the Malampuzha project in Kerala, the irrigation potential was found to have increased to 83 percent compared with 70 to 80 percent earlier after the new management practices were implemented.

"In conclusion, several case studies in South India have shown that satellite remote sensing techniques can provide reliable and objective primary data on agricultural productivity and secondary data on related water management aspects to help better water management in existing irrigation schemes." Cost-effectiveness and time-effectiveness is a major advantage in using satellite remote sensing data and integrating several other data systems using computerised GIS for taking rapid management decisions. With the new IRS-1C data, it is expected that much more can be done to help increase agricultural productivity and more efficient water usage in various existing command areas.

Some of the measures that need to be taken more effectively are:

a) In situ conservation Practice

b) Contour bunding

c) Construction of check dams at appropriate places

d) Forming farm ponds in individual farmers' fields

e) Construction of percolation ponds

f) Construction of irrigation tanks

g) Utilisation of groundwater properly keeping in view recharge capacity

h) Resource management

i) Alternative land use, etc.

In all these cases, the use of remote sensing will be a versatile tool to speed up development activity.

27

Satellites and Aquaculture

Food production is declining—the most severely hit sector is fisheries. A viable alternative is to develop aquaculture facilities in a planned manner using remote sensed data and a GIS to select sites for aquaculture and to have the least possible adverse effect on the coastal environment.

Experts say that, globally, the ratio of growth of food production to population growth has entered a period of decline over the past decade. This includes all types of stable crops, meat, etc. However, in the case of fish, there has been a net decline of seven percent in per capita world production. The aim of this article is not to examine the reasons for this trend, but only to focus attention on the limited aspects of brackish water aquaculture and an effective way of site selection that could lead to better production without much adverse ecological impact. This write-up is based on the study and the report by the scientists of the Marine and water resources division of Space Application Centre (ISRO) of Ahmedabad, in 1995.

The study was supported with funds by the Ministry of Agriculture of the Government of India and the Marine Products Exports Development Authority of Gujarat. The use of satellite data and GIS have been found to be advantageous tools both for initial selection of proper sites as well as for continuous monitoring (Plate 27.1).

The world fishery production increased from 3.5 million tonnes in 1956, to 76 million tonnes in 1975, and subsequently the rate of increase in production was not to the same extent as already pointed out. The reasons could be many. The emphasis has been to increase both brackish and inland water fisheries. It is in this context that a study was perhaps undertaken by the Space Application Centre to use space technology combined with GIS to streamline the increased production activities through proper site selection and regular monitoring.

The coastal ecosystem is fragile and any activity such as intensive brackish water aquaculture, however beneficial it may be to the rural poor, may pose an environmental threat and also affect sustainability. Using remotely sensed data, a baseline land use information was in this case generated initially for all maritime states, and using this and various other parameters and in consultation with appropriate authorities, the scientists of the Space Application Centre developed two user-friendly computerised GIS, which were found to be highly beneficial. To arrive at this GIS software, they had used a total of 29 parameters, to retrieve and prioritise areas most suitable for aquacultural development, which will have least adverse impact. Without the use of appropriate remotely sensed data and the GIS, it would not have been possible to arrive at suitable solutions.

The information needed for such a study was in the form of producing initially thematic maps such as land use, soil type and water quality; obtaining engineering, ecological, meteorological and infrastructural and socio-economic data. Within these main information needs, were several detailed analysis of such data more quantitatively analysed wherever possible, so that these could be integrated and used as an input to the GIS processing through the use of computers. The GIS is a powerful tool for collecting, storing, organising, retrieving, transforming and displaying to aid taking the most appropriate decisions for any specific purpose.

Besides preparing land use maps on a 1:50 000 scale using the Indian Remote Sensing satellite data, LISS II, some data from the US Landsat and the SPOT, the Survey of India topographical maps in 1:50 000 scale were also used to a generate proper cartographic database to enable accurate positional needs.

It may not be necessary to go into the details of various operations carried out to avoid technicalities, as the aim of this article is to highlight the application possibility of use of remotely sensed data along with a suitably created GIS package for obtaining the maximum benefit.

With the help of better data from the recently launched IRS-1C satellite, such studies could enhance the scope by greater detail and precision, to evolve a technically feasible, economically viable and socially acceptable planning of improved coastal area management and contribute to increased food availability.

28

Using Land through Remote Sensing

With the intention of evolving a Land Use Policy, the Government of India set up State Land Use Boards in each state to determine and monitor land use, using remote sensing as a tool. This was aimed at optimal land management to feed our growing population.

Diversity of the physical landscape is the cumulative effect of both geological and geomorphological processes of several million years. A variety of land systems such as mountains, hills, valleys, plateaus, plains have been formed. It was land and the water which sustained human and other living beings. While the population was low, per capita availability of land, whether cultivable or otherwise, was able to withstand exploitation and there was no adverse impact. However, over a period of time, with the rapid increase in population, per capita availability of land declined and proper and planned utilisation of land became necessary for our survival and sustained human existence. Over-utilisation of land or improper agricultural practices such as overgrazing, exploitation of forests, etc. resulted in undue pressure and degradation. In fact, the per capita availability of land in India declined from 0.9 ha in 1951 to 0.5 ha in 1980–81. By 2000 AD, it is likely to decline further to 0.15 ha.

To increase productivity, traditional subsistence farming has given way to intensive agricultural practices which have in turn, given rise to problems like increasing salinity, soil erosion, water logging and general degradation of land. Better measures for land use will not only ensure sustainability over a longer period of time but also adequate production to feed the ever-increasing population.

The Government of India, in 1974, decided to evolve a national land use policy for managing land optimally. The most efficient use of this resource was to be determined on the basis of surveys relating to land capability and technology of production, and the policy framework was to be organised so that land was used for suitable purposes on the basis of efficiency of production.

The Government of India requested the states to set up boards to determine and monitor appropriate land use in an integrated manner. The National Land Use Board was also set up in 1983 with the same objective. While the objectives of these boards were laudable, considerable information about land and mechanisms for implementation were required but were not available. Even today, the land use boards may exist, but how far they are efficient is questionable, as till recently most such boards were unable to obtain real-time data and did not use modern technological tools. It is in this that space technology combined with computer technology provided an effective tool to bring out a variety of information needed to implement a sustainable land use planning activity. There are several activities connected with such an operation, including planning on the basis of watersheds and even micro-watersheds based information system, which is now available through satellite based remote sensing, and the use of GIS to enable appropriate decisions and also ensure the least interference from the environmental point of view.

Land use is referred to "Man's activities and various uses which are carried out on land," and land cover is referred to "natural vegetation, water bodies, rock/soil, artificial cover and others noticed on the land." Since both land use/land cover are closely related and are not mutually exclusive, they are interchangeable as the former can be inferred based on the land cover and on the contextual evidence.

Having realised that land use/land cover information is the basic prerequisite for land, water and vegetation resource utilisation, conservation and management and also that such information is required periodically to monitor changes, particularly in areas where such changes occur more frequently, it became necessary to use technology and tools like remote sensing from satellite altitude.

Now that remotely sensed data from satellite altitudes has become almost an operational technique, and realising the need for an up-to-date and accurate land use/land cover maps especially for effective agricultural land use planning and at the behest of the Planning Commission, Government of India, the NRSA, under the Department of Space, undertook and carried out land use/land cover mapping for the fifteen Agro-Climatic Zones (ACZ) covering all the 442 districts in the country on a scale of 1:250 000, using satellite data pertaining to two seasons (Kharif and Rabi). The Indian Remote Sensing Satellite LISS I (nearly 76 m ground resolution) of 1988-89 was used for this exercise (Plates. 28.1, 28.2, 28.3). In this case 22 land use classes, which were developed by the NRSA, after discussion with various departments were finally depicted in the maps. While the NRSA coordinated the entire effort, a number of other organisations such as State Remote Sensing Application Centres, the Regional Remote Sensing Service Centre (RRSSC), the Central Arid Zone Research Institute, the Centre for Studies in Resources Engineering of IIT Bombay, and the Space Application Centre of ISRO, participated, sharing the work load to complete the task expeditiously. The final reports are also being distributed on request to a variety of institutions, including NGOs.

The salient feature of the methodology adopted consisted of using visual interpretation techniques for 274 districts, and 168 districts were mapped using digital techniques and computers with appropriate software. The land use/land cover area statistics derived from satellite data at the district level for the base year 1988-89 were compared with the figures published by the Bureau of Economics and Statistics (BES) of the Ministry of Agriculture (MoA), Government of India, for the same period for consistency and compatibility, and reconciled wherever a discrepancy was noticed. Using remote sensing techniques for mapping, monitoring and managing land uses and also planning fresh land uses is highly advantageous and cost-and time-effective with very reasonable accuracy limits. With the new IRS-1C data which has now become available, with 24 metre ground resolution of multi-spectral data (colour) and at 5.8 m ground resolution in black and white form (panchromatic) mapping at a larger scale, say 1:50 000 or even a larger scale in some cases, it will be possible to get more accurate information. Hopefully, the State Land Use Boards, will accept this method to obtain periodic information of a variety of land uses on a regular basis.

29

Watersheds and Remote Sensing

Remote sensing data is cheaper than aerial photography in covering vast areas. One means of water management is through the creation of watersheds. This leads to increase in ground water and vegetation as evidenced by the Kalyanakere-Mavathurkere watershed project in Karnataka. To be effective, the approach must be participatory by involving the local beneficiaries.

Before the advent of remote sensing data from satellites started arriving in the early Seventies, the use of aerial photographs for surveying and mapping natural resources was well recognised, though generally based on manual interpretation. Aerial photographs were first used for military reconnaissance and later for the production of topographic maps. Their usefulness in natural resources mapping was highlighted during the early Sixties, specifically at an international conference held in Toulouse, France, which a few of the senior officers of the Survey of India attended.

Realising the need for such maps in India, the Indian Photo Interpretation Institute was started in Dehra Dun. It is now known as the Indian Institute of Remote Sensing and works closely with the Institute for Aerospace Survey and Earth Sciences of the Netherlands and the Survey of India. In the early Seventies when remotely sensed data from satellite platforms was collected and when the National Remote Sensing Agency was formed in 1975, it was thought appropriate to make the Indian Photo Interpretation Institute the main training centre in this field. While aerial photographs could be used to study in three-dimensional form, using instruments like stereoscopes, satellite data had to be interpreted or studied monoscopically. But the multi-spectral colour images that could easily be obtained from satellite data were a great advantage and highly economical because it could cover a large area in one picture and a picture of the same area could be obtained frequently, which if it was to be done by aerial coverage would be very expensive.

All such tools or data enabled the study and mapping of natural resources and its potential development to plan better and thereby get more out of land than what was possible before. As the holdings of most Indian farmers were small, the increased productivity had to be undertaken at a micro-level, rather than at macro-level. While several environmentalists and scientists warn us about the serious land degradation that has been taking place due to certain intensive agricultural practices, it is commendable that India was able to increase its food production almost four-fold during the past half century. It is the belief of one school of thought that, with a variety of newer R & D activities and farming techniques, India will be able to increase its food productivity to keep pace with the needs of the increasing population, without much damage to the land and on a sustainable basis.

It is in this context that this article is being written—to highlight "watersheds" as a basis for more effective land management, by adopting a participatory approach of the local people. In this context it may be necessary to define what a watershed is as many people may not have a clear idea about this term. The Dictionary defines watersheds as "line of separation between waters flowing to different river basins or seas," "A turning point in water flow or ridge of high ground." This need not necessarily mean only large river systems; even small streams can be considered.

Once a proper assessment of such small watershed areas is made, it becomes possible to use water optimally for agricultural operations by small farmers, where water is one of the main commodities. Such smaller watershed delineations becomes possible with better or improved resolution satellite data, integrated with ancilliary information. While several reports might have been made in the past by experts using remote sensing, a participatory approach using the co-operation of local people who are the beneficiaries has emerged recently.

It is no use if reports are made and presented to high officials in colourful functions. Its active usage needs the participation of the actual beneficiaries, who should convinced the idea. This approach will fructify only if people find it rewarding economically. The use of remote sensing is not a panacea for all ills or all problems, but it is a very effective tool.

Recently, the main remote sensing organisations like the NRSA and the state remote sensing centres have embarked on a plan of integrated resource development programme in a number of districts, to enable sustainable development. How far such a report is translated into usable form by the beneficiaries is not yet very clear.

It is in this context, that I came across one report "Towards Sustainable Resource Development", the experience of Kalyanakere-Mavathurkere Watershed in Karnataka, where they have also used remotely sensed data provided by the regional remote sensing service centre of the Department of Space, Bangalore. According to this report:

"Watershed discipline" has become a way of life in that region. This publication has stated it "makes a very detailed inventory of the technology and practices followed in the watershed and the results obtained. It shows graphically how the people were convinced to change their own lives." (Plate 29.1)

The programme was initiated and successfully implemented by the Directorate of the State Watershed Development Programme. It must be emphasised here that remote sensing data was only used as a part of a variety of efforts that were undertaken. In fact, this watershed development programme highlighted two aspects: one, a sustainable increase in the agricultural productivity and two, making dryland farming a viable enterprise, with the people's participation.

The success attained over a period of 8–10 years is a clear indication that similar practices, suitably modified if necessary, based on soil and other land capability parameters can enable increased productivity in areas where individual holding is small and proper delineation of watershed and sub-watersheds can be done efficiently (Plate 29.2).

The watershed delineation was done using topograhical maps, revenue maps, etc. However, if a more detailed topographical map of a larger scale could be made available it would make planning at the micro-level better.

The Indian Remote Sensing Satellite data pertaining to November 1988 and November 1993 (a five-year gap period) were used to make an assessment of change, by the Regional Remote Sensing Service Centre at Bangalore which clearly showed the improvement that has occurred in vegetation coverage both in respect of area and vigour and also in respect of increased small water bodies.

In conclusion, it was evident that increased cultivation, more number of water bodies, changes in the surface water spread area, increased area under tree cover, etc. can be attributed to the participatory watershed development programme. Now with the availability of improved satellite remote sensed data, it is felt that a much better and more quantitative assessment of various developmental activities on the Earth's surface should be possible to help planners and others to decide on future courses of a variety of actions. This could be an example of the saying, "Think globally but act locally."

30

Resolutions in Remote Sensing

Higher resolution remote sensing can give us a better idea of the world's resources and could lead to a more sophisticated Land Information System (LIS). The author suggests that the Indian authorities review the classified documents act which is outdated and make information more freely available.

Some aspects that are important to satellite imagery, are spectral resolution, spatial resolution, temporal resolution and radiometric resolution. Spectral resolution defines the areas in which a particular band or bands bring out characteristics exhibited by certain features on the surface of the Earth known as spectral signatures. For example, objects on the surface of the Earth, which are predominantly green, say forest cover and water bodies, have their own specific or reflective characteristics. The narrower the band in which we try to record features on the Earth, the sharper the distinctions. However it is not possible—or economical—to have too many spectral bands in one sensor. So the most reasonable values are combined and the entire data recorded in three or four spectral bands.

In some cases, one or two in the infrared region, thermal region or microwave region of the electromagnetic spectrum may be added. The explanation may be a bit confusing to those not familiar with the topic. But it is enough to know that the spectral band or the spectral resolution which separates certain regions of the electromagnetic spectrum is one of the parameters which has to be taken into account in designing sensors, to enable maximum utility of the product to the users.

As far as temporal resolution is concerned, data regularly obtained of the same area at periodical intervals gives us the facility of detecting change. A satellite like IRS-1A and -1B repeats their visits to the same area, once in 22 days and with two such satellites simultaneously in orbit we get data once in every 11 days. When we consider the use of other data for satellites put in orbit by other countries, we may be able to get information more frequently if we have access to them.

Talking about the spatial resolution of the size of the object that could be discerned or identified on the ground with data purely on the panchromatic (black and white) mode in a single visible band, it becomes easier for the users to carry out more a detailed study of features to differentiate in terms of its dimensions. If we say 2 m spatial resolution, it means that it is possible to decipher objects which are 2 m in length and breadth.

Till recently, except for the French satellite SPOT where we had a 10 m spatial resolution, panchromatic data that was freely available, although at a higher cost. This could be merged with other multi-spectral data of say 30 m resolution, like the US Landsat TM or even IRS-1B LISS II data (35 m), using suitable software. Now that the IRS-1C has become operational, we can get PAN (Black and White) data similar to the French SPOT 10 m data, but with 5.8 m resolution by which the utility, interpretability and intensity of usage has improved tremendously. As a matter of fact, the quality of our IRS-1C data has a 5.8 m ground spatial resolution; it took some time to get such data for the public, perhaps due to the intervention of security organisations. Now this information is freely available and people are investigating how best it could be used for many applications in several areas.

At present, the Survey of India 1:50 000 scale or 1:25 000 scale topographical maps are restricted to private organisations, falling 50 km along the coastline and along international borders. Under these circumstances, what will be the reaction if satellite data of 1 m and 2 m ground resolution start becoming available for purchase from other countries covering any part of the world such as from the US and Russia in the near future, from which much more information could be obtained? How will our security agencies react to their usage? What will be their concern, if minute details can be identified and extracted from such data?

Already some parts of India are covered by the Russian 2 m resolution data and are available for purchase outside India. There is data from German cameras of recent origin called MOMS (Modular Optical Mechanical Sensor) camera available with about 4 m resolution. The US Government has become more liberal today and has allowed private companies to launch their own satellites to acquire 1–3 m resolution data which, in some cases, could even be stereo in character. Such data perhaps could be very effectively used even for mapping our cadastral land holdings which are small in nature, say about 1–3 ha in general and could lead to the production of a more sophisticated Land Information System (LIS) combined with data using Global Positioning System (GPS). The reaction of our intelligence

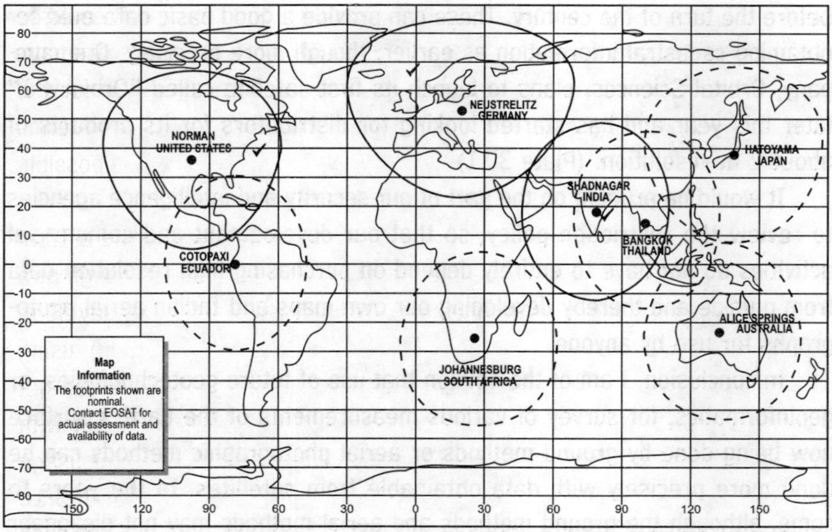

Fig. 30.1 *IRS-1C Network of international ground stations*

agencies both defence as well as civil is not clear. (Fig. 30.1).

Unfortunately, the British regime some 50 years ago left a legacy of an Official Secrets Act under which maps and aerial photographs came under the classified category. There were maps meant for use by the 'Defence forces only, in certain localities and to enforce the aspect of the Official Secrets Act much money is spent on a large workforce'. In view of the changed circumstances and highly sophisticated technologies having been evolved, is it worthwhile or even desirable to enforce such severe security restrictions?

Certain areas and objects or points may be located only in certain areas in our country and these alone could be guarded. For quite some time now, security restrictions on maps and aerial photographs and their non-availability for genuine and bonafide use by private agencies and the public have been under severe critical debate. But no one seems willing to boldly decide to relax these restrictions. Every country has its own methods of guarding vital information in some of the maps, defence or civil. But changes have to be commensurate with technological developments and actual necessity.

It is also clear now that the operational satellite system is no longer exclusively launched by governmental agencies in the US and a few other countries. Commercial companies have come forward to launch high resolution satellites covering the world. In the US alone there are eight or nine companies which may be launching such high resolution (1–3 m) satellites before the turn of the century. These can provide a good basic data even for obtaining cadastral information as earlier, though more precisely. One company, Orbital Sciences, plans to launch its first satellite called "Orbview-1" later this year and has started looking for distributors for its products of about 2 m resolution. (Plate 30.1)

It would be prudent on the part of our security and intelligence agencies to review the restriction policy, so that our development and commercial activities do not have to entirely depend on purchasing high resolution data from outside and thereby developing our own maps and Indian aerial photographs for use by anyone.

In conclusion, I am of the opinion that use of future geotechnologies, or geoinformatics, for survey of various measurements of the Earth's surface now being done by ground methods or aerial photographic methods can be done more precisely with data obtainable from satellites. In the years to come, although the ground methods and aerial methods may not disappear totally, the future appears quite encouraging with the advancement being made in space technology.

31

Gaining Ground Rapidly

This brief describes how the application of remote sensing can help raise productivity of foodgrains appreciably, citing how beneficial it has already proved to be to the fisheries sector.

Agricultural activity in the country can be considered a large private industry. Although the government helps farmers in their efforts to increase production by providing various types of information and also arranging for irrigation facilities, making available fertilizers and other inputs, it is the farmers who own the land and do the actual farming, who really matter as they can only be persuaded or provided with incentives to improve productivity. The same is the case with fishing operations. Over a period of four decades foodgrains production has risen from about 50 million tonnes to about 170 million tonnes in 1990.

There are several reasons for this growth. There was a time in the early Sixties, when several western countries predicted that India would be faced with serious famine conditions and that hunger would take a heavy toll of human lives. This prediction proved false. The Green Revolution changed the scenario rapidly to one of self-sufficiency. One cannot therefore deny that several scientific and technological inputs have also played their part in the efforts to increase production of grains as well as fish.

In the area of shrimp and prawn export, the corporate sector has stepped in. Similarly, can a strategy to bring grain production under the organised corporate sector be thought of? There is also the fact that the per capita area under agriculture is getting reduced, due to increasing population. One of the important inputs in raising farm productivity will be up-to-date and relevant information, on a near real-time basis, based on a technology like remote sensing. While considerable work has been done using remote sensing in the area of agriculture, the achievements during the last one to two years in the application of remote sensing from space can be reviewed.

What is Remote Sensing?

Remote sensing is the science of collection of information about any object without physical contact. The advantages of remote sensing are:

i) Synoptic views—A single image of approximately 23 cm × 23 cm format size can cover a very large area on the Earth's surface. For example, some of the satellite remote sensing imagery can cover about 30 000 sq km in just one scene.

ii) Repetivity—Satellite remote sensing data can be obtained frequently of any part of the Earth's surface. The revisit ranges from about once in 16 days to about once in 26 days, with respect to different satellites. If the usage of multi-satellite data is assumed, one can get data as frequently as once a week or even less subject to certain conditions.

iii) Such satellite remote sensed data can be used to study the natural processes taking place in areas to which there is no easy access.

iv) Satellite remote sensed data are available in both digital and picture format. The digital format enables using such data for processing or interpreting through computers.

v) Remote sensed data till recently were available only when there was no cloud cover. But now microwave remote sensing has eliminated this problem. During the last one year or so, the European Remote Sensing Satellite has produced data by using microwave sensors, and clouds do not pose a serious obstruction.

As for ground resolution obtainable in satellite remote sensing, it is now 10 m in the case of the French SPOT satellite, though mostly the data obtained from the Indian Remote Sensing Satellite (IRS) is being used in the country.

With two similar Indian Remote Sensing satellites in orbit now, the country can get data from its own satellite almost every 11 days, though the ground resolution is of the order of 36 m and 72 m. In fact, for most applications, these data have been found quite adequate. It is also expected that in about two years from now, the country is likely to get a Remote Sensing Satellite of a new generation which will give a ground resolution greater than 10 m and with the possibility of stereocoverage.

In the last couple of years, considerable progress has been made in the areas of agriculture, land use mapping, mapping areas under saline oralkaline conditions, the study of drought conditions using weather satellite data, near operationalisation of determining sea surface temperature, identification of erosion prone areas in selected major watersheds, besides mapping and monitoring of forest/vegetative cover in the country.

Its Applications to Agriculture

The use of remote sensing data for estimating crop acreage estimation has reached a stage of near operational level. Several studies carried out for estimating acreage under different crops in different parts of the country show a near 90 percent accuracy level. Experimental work covering the rice crop in Tamil Nadu, jointly by the Space Application Centre of ISRO, the Institute of Remote Sensing of the Anna University and the Directorate of Agriculture of the Tamil Nadu Government has shown very encouraging results. Similarly, wheat acreage estimations covering some of the Northern States have proved highly successful. In such studies, digital analysis based on powerful computer systems has been found useful (Plate 31.1 (a) and (b)).

Yield Forecasting

Yield is influenced by many factors, such as crop genotype, soil characteristics, cultural practices adopted, meteorological conditions, and influences of diseases and pests. Many approaches have been followed to determine the integrated effects of various parameters that affect crop growth and yield. Several yield models have been developed in which data obtained from various types of satellites to cover some of the parameters have been used. In the case of wheat in Punjab and Haryana, the use of simple models obtained from different types of satellites has led to very good results in some individual districts, deviating only to the extent of eight to 10 percent.

In the case of rice, where 18 percent of the total production in the country is obtained from Tamil Nadu and Orissa, the study has shown reasonably good results but some more work remains to be done to reach an operational level. Similarly, studies are continuing in the case of sorghum in Maharashtra and oilseed crops in Gujarat.

To improve upon the existing methods of collecting statistics and information in respect of area, yield and plant disease/pest, it has been proposed to use space technology within the framework of the existing agricultural information system without disturbing its hierarchy or organisational structure. The idea is to create an ability to constantly upgrade itself by absorbing the latest in technology. It has also been proposed by the Department of Space, that each state should have one image processing facility to use satellite remote sensed data for analysis on as near real-time basis as possible. It is expected that by using appropriate sampling techniques, it should be

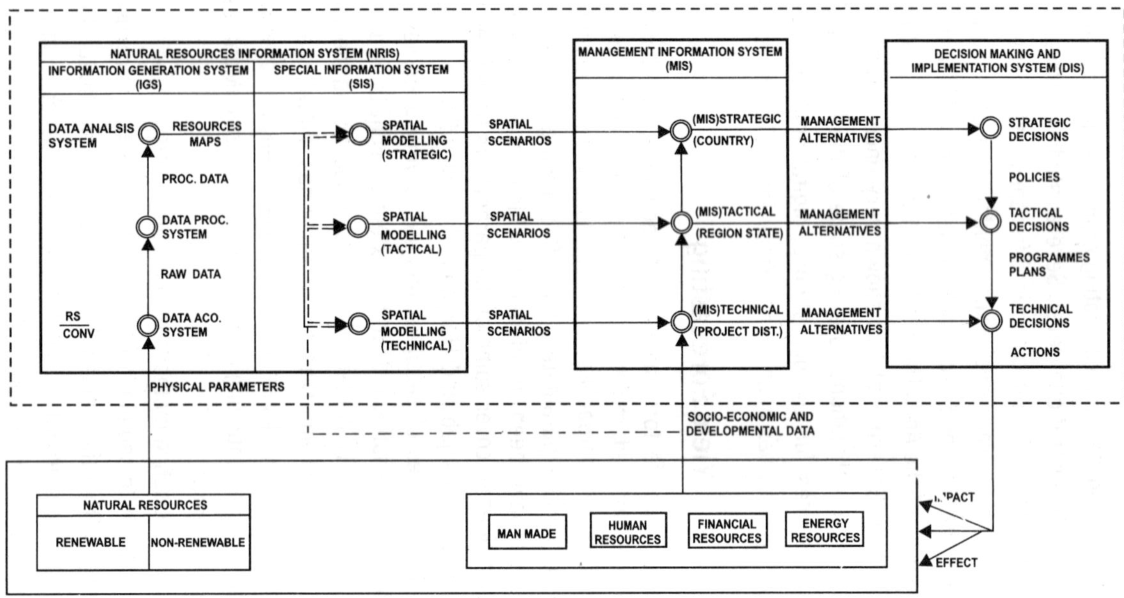

Fig. 31.1 *Framework of Natural Resources Information System (NRIS) (Source : SAC–ISRO)*

possible to arrive at crop production forecasts to a reasonable level of accuracy. Inputs from metrological departments on weather information will also be necessary. In fact, proper integration of various techniques and methods of data gathering can lead to better crop estimation.

It may be mentioned here that remote sensing techniques have been operationally used in the country to provide basic information on crops, soils, water resources, and the impact of drought and flood on agriculture. Integrated studies on soil and water conservation using remote sensing and GIS have been progressing with a view to raising agricultural production. Procedures for pre-harvest acreage estimation of major crops such as wheat, rice and sorghum, using sampling and digital techniques based on remotely sensed data have improved greatly.

Fisheries

During the past few years, there have been intensive activities in the use of remote sensing for ocean applications, in which obtaining sea surface temperature from advanced very high resolution (AVHRR) data from weather satellites, forms an important component. The Department of Ocean Development gave a thrust to this activity and the provision of this information to fishing co-operatives and fishermen communities had helped increase production.

The study of ocean colour and the sea surface temperature have proved to be valuable tools for establishing a relationship between fish distribution (particularly tuna fish) and its availability. The Space Application Centre of ISRO had developed some methodologies to use remote sensing technology to aid the location of fishery resources. This work was being carried out with the Central Marine Research Institute, Kochi, and the Fishery Survey of India, Bombay. To support the satellite data, several sea truth measurements were also undertaken for correlation. The results have been highly encouraging.

During the past one year, the NRSA has been developing sea surface temperature information using weather satellite data and disseminating this to the fishermen community through various media as well as by fax and telex, wherever possible.

The response has been very encouraging from the user community. In fact, there have been considerable demands now from some users such as the Gangolli Fishermen's Cooperative Society in Dakshina Karnataka, for such data through telephone for saving diesel consumption and improving the rate of production. There are several such requests to the Director of the NRSA to supply sea-charts more frequently through various communication systems available. There have also been demands on the Department of Ocean Development for installation of fax facilities in various coastal areas to enable speedy dissemination of processed data from the NRSA. It is reported that in some cases the fishermen community has been able to catch four to five times more than their normal level by using such information obtained by processing data culled from satellite platforms. (Fig. 31.1)

It can well be said that information is power and if it is disseminated quickly to those who need it, then it brings in more economic benefits. Studies are now being conducted by various institutions (for example, the M S Swaminathan Research Foundation, Madras) to devise a methodology by linking the information creators and information users through suitable multi-media techniques, to convey it to the last man on the ground, as it has been established that with more information, productivity increases. This applies equally to the agricultural sector.

An interdisciplinary dialogue organised on the topic of "Information Sciences & Technology—Reaching the Unreached" by the M S Swaminathan Research Foundation in January 1992 has come out with various suggestions that include creating information villages which will contribute to improved agricultural activities and consequent improved production. Remote sensing technology will have a crucial role to play in this effort. It will also be highly beneficial if private entrepreneurs are trained in the use of remote sensing, as well as in the use of rapid methods of dissemination of information to the person at the bottom-most level.

32

Valuable Aid for Crop Production

While every factor causing land degradation cannot be monitored by remote sensing as of now, certain important factors can be identified periodically from remote sensed data.

The area under agriculture in India is about 144 million ha according to official figures. But the NRSA which surveyed and mapped land use/land cover, puts the total land at about 20 percent more; this gives rise to doubts about the productivity figures announced from time to time. This higher figure could be due to agricultural activities on lands belonging to the Railways, along the tracks, land on the side of some roads and encroachment on forests. Or there may be an error in the interpretation of the satellite data, and consequent misclassification due to inadequate ground truth inputs.

In 1990 or so, the NRSA, undertook the mapping of land use/land cover using Indian remote sensing satellite data products, to obtain maps on a 1:250 000 scale. There was a total of 375 such maps covering India. With such a small-scale map and using satellite data of a spatial resolution of around 75 m, small patches of land and their content cannot be easily mapped. A size of 3 mm × 3 mm area on this scale will cover about 56 ha. Secondly, there could also have been a certain amount of misclassification because of inadequate ground checks. Nearly half of this land use/land cover classification was attempted by digital techniques using computers and the other half by visual interpretation. If the remote sensed data had been produced after enhancements most suitable for such land use/land cover mapping, perhaps visually interpreted results would have improved considerably.

On the other hand, for digital techniques an adequate number of so-called "training sets" collected from the ground had to be fed in, for the computer to recognise different land use classes correctly. This could also create errors. However, in both cases, an overall level of accuracy could be around 80 to 90 percent and, therefore, the total area estimated under any one category of land use could also generally match this figure.

The question that arises is in what respect can remote sensing be used in agricultural information systems. In fact, agricultural production encompasses myriad activities related to crops, soils, water resources, climate and local weather of the region. For some of the specific themes of these disciplines, remote sensing technology has been observed to be operational and in some others some more research may be needed to enable integration into the total agricultural information system.

Mainly, an agricultural resources inventory may need some knowledge of geology and geomorphology, forestry and vegetation, land use, soil use classification and information, agricultural crops, water resources, agrometeorology, and land degradation.

The use of remote sensing data to obtain information and produce maps on a 1:50 000 scale is almost operational in all the aspects as mentioned above. In the case of geology and geomorphology, there are many examples of work done covering many parts of India successfully.

Forestry and vegetation forest cover and its status are being mapped regularly by the Forest Survey of India, once in about two years to monitor changes. In specific areas, even forest species have been mapped on a scale of 1:50 000 using Indian Remote Sensing Satellite data, successfully.

Regarding land use, as mentioned above, all of India has been covered on a scale of 1:250 000 using IRS-1 A data, and in specific regions more detailed information can be obtained on a scale of 1:50 000 as of now.

As to soils and soil classification maps, it is gathered that detailed soil mapping is a long process and only around 30 percent of India has been covered. However, the National Bureau of Soil Survey and Land Use Planning has covered the entire country on a 1:1 scale and in some parts even on a 1:250 000 scale survey. How far a more detailed coverage on 1:50 000 scale has been completed is not known.

For sustainable agriculture and production, cropping patterns, soils, water resources and climatic conditions have to be studied in detail. A substantial area under agriculture depends on rainfall which is erratic and unpredictable. The excess of precipitation that occurs in many parts needs to be effectively conserved and properly managed to ensure availability of moisture for a longer duration. As far as soil is concerned, the process of degradation has been quite serious. Hence, a realistic monitoring procedure is essential to ensure timely corrective measures, otherwise a time may come when such degraded areas cannot be reversed, for production activities. Remote sensing as a tool will give information regularly and at a low cost to enable timely reclamation.

One of the first exercises carried out in NRSA in 1978 was to make an assessment and locate areas affected by salinity in the entire State of Haryana using data got from Landsat 1 and 3, with a spatial resolution of about 80 m. The work took about eight months and a report was prepared, identifying areas highly affected by salinity, areas of medium salinity and areas of very moderate salinity. Since the spectral responses or reflected characteristics are very specific as seen by salt incrustation on the surface, identification was not difficult and the Haryana Government made use of this information to reclaim the area and increase productivity. While the information could only be qualitative in nature at that time, a more detailed and quantitative estimation would certainly be possible using better quality satellite data. These are becoming available and will be a regular data source by the turn of the century.

A report from the World Resources Institute has pointed out that some 1.20 billion ha has been seriously degraded since 1945, all over the world. Many reasons have been attributed. Excessive use of certain agricultural practices has only led to a speedier deterioration of the natural resource base, very essential for sustained agricultural practices with consistent production capacities.

Erosion, salinisation, compaction and depletion of organic matter have made an estimated 2 billion ha of once-arable land irreversibly unproductive, says the World Resources Institute in one of its publications.

While every factor causing degradation cannot be monitored by remote sensing as of now, there are certain important aspects that can be found out periodically from remote sensed data.

Dr M S Swaminathan says the new frontiers of technologies that may help in maintaining the level of production and also lead to sustainability aspects are biotechnology, space technology including conservation monitoring, information technology, micro-electronics and management technology. Space technology in the form of remote sensing applications and information technology are inter-related for the purpose of land and water management activities on a continuous basis and provide systems to support management decisions. However, in this article, I will only attempt to describe how remote sensing can be used in mapping and monitoring land and soil degradation and also for crop inventory assessment (Plate 32.1).

Remote sensing application for mapping, monitoring and management in India is considered well advanced, and comparable to many advanced countries. With the launch of the next generation remote sensing satellite called IRS-1C, we will be able to have 6 or 7 m spatial resolution black and white panchromatic form, a 24 m spatial resolution in multi-colour and the possibility of viewing stereoscopically. That will give us an idea of the height information within limits. Further, a sensor called WIFS will be able to cover in swath a much larger area, very useful to monitor ocean parameters also. This satellite is expected to be launched by the end of this year and there are enquiries from some of the advanced countries to tie up with India for getting these data, directly or through on-board recorders.

Precise information on the nature, extent, spatial distribution and magnitude of the problems in degraded lands forms a vital input with conservation, reclamation and scientific management of these land and try to reclaim as much as possible by suitable treatment to activate economic agricultural activities. It is also important to monitor these degraded lands at regular intervals to know the impact of the reclamation and conservation measures, possible now, as remote sensed data is available very frequently, say from once in about 10 days to once in about 26 days, using multi-satellite data. Interpretation of remote sensed data can be carried out either visually or digitally, using computers, as the original data is available in digital format. Good quality and well enhanced pictures are probably very good for visual interpretation to obtain almost 70 percent of the required information, and will be cheaper in countries like India, where cheap labour is available in plenty.

Remote sensing data can help locate soil erosion, areas under shifting cultivation, areas severely affected by salinity and areas covered by ravines. The scale of mapping is almost operational at 1:50 000 and probably when new IRS-1C data start coming in the scale can extend to 1:25 000 or even larger.

In the case of waterlogged and marshy areas, remote sensing data has given highly successful results (Plate 32.2). Other activities include watershed priorities, desertification, land affected by mining and land affected by large-scale inferior coal-based power generation.

It is stated that Indian farmers generally use 20 percent more water than actually needed. Due to large-scale dams and canal systems and water seepage, land gets degraded very soon and allowing the good lands to remain idle for a long time turns them into wastelands, which are quite extensive in India. Excessive extraction of groundwater without adequate recharge leads to ground-water level going down sharply. One reason could be use of high power electric pumps and the government encouraging the use of such pumps by providing free power and loans at low rates of interest. While such measures give short term relief, in the long run it becomes unsustainable.

Another important use of remote sensing data is for crop production or crop area estimation. In fact this was one of the first attempts made by the US in 1974-78, called LACIE (Large Area Crop Inventory Experiment) to monitor wheat growing areas around the world using remote sensed data. This experiment was good for some large areas like the former Soviet Union, but found inadequate for certain other countries such as China, Australia, Brazil, India and Argentina.

Agriculture and Resource Inventory Surveys through Aerospace Remote Sensing (AGRISTARS) was initiated in 1980 by the US mainly to meet the needs of production information.

In India, production forecasting of certain crops, crop yield modelling and crop stress detection are done using remote sensing data. Crop acreage estimation has been quite successful, through both visual and digital analysis in many areas. A major constraint however is a cloudy sky during the cropping season when normal optical remote sensing cannot give good data. However, with the emergence of microwave remote sensed data, this can be overcome as such instruments can penetrate through clouds. To ensure complete success in predicting yields, some more research experiments may be needed to evolve a foolproof system and the Space Applications Centre of ISRO is very active in this area.

Certain agencies like the Tamil Nadu Agricultural Department have shown some interest in such an activity together with the Space Applications Centre of ISRO and the Institute of Remote Sensing of Anna University in respect of rice, sugarcane, cotton and some other crops.

Various factors such as intensive cultivation, monocropping, changing weather conditions and indiscriminate use of pesticides have resulted in frequent outbreaks of crop pests and diseases causing huge crop losses. Minimising these losses is one way of enhancing grain production, and remote sensing as a tool has been found very useful in monitoring large areas frequently. The Earth observing systems are useful in monitoring weather and ecological conditions favourable for crop pests and diseases.

An aircraft fitted with a camera loaded with colour infrared films and flown over Kerala, identified coconut areas severely affected by 'wilt' which could not be easily detected from the ground. This gave a clue to the area that could be viewed from satellite altitude.

Understanding the magnitude of crop losses is necessary to appreciate the importance of plant protection in crop production programmes. Losses can be due to biotic factors such as pests/diseases/weeds and abiotic factors such as drought, floods, cyclones and hailstorms. Damage caused by pests may be quantitative (overall reduction in yield), or qualitative such as change in colour and offensive odour. The regional disparities in crop condition assessment, the complex Centre-State relationships in handling relief measures and the introduction of crop insurance scheme, call for an unbiased, objective and timely information system to give early warning, to indicate the intensity of such hazards and to assess the loss.

The FAO had organised an international training programme at NRSA a decade ago, involving countries like India, Pakistan and Tanzania in the use of various remote sensing data to identify locust-breeding areas. The directorate of plant protection handles such issues. There is no proper information about the area affected by pests and diseases and other yield reducing factors on an all-India basis.

It is stated that the brown plant hopper (BHP) of rice is one of the dreaded insect pests in Asia. BHP is stated to be associated with synoptic weather conditions (depressions). Double cropping, extensive rice cultivation in the command area and indiscriminate use of fertilizers and pesticides aid the occurrence of BHP. Much better quality satellite data from the new IRS-IC satellite expected to be launched in the coming months will certainly be very useful.

To sum up, the use of satellite based remote sensing has proved itself a strong and unbiased information system at regular intervals of time. While agricultural scientists have shown some interest in developing its usage, there is still a long way to go, as it is only the agricultural scientists who can clearly define what information is actually needed. Besides, they should integrate the remote sensed information system with their agricultural information system to derive optimum usage, timely recovery of degraded land and refrain from unsustainable activities by use of other advanced technologies to their benefit and to enable increasing productivity through alternate farming systems.

33

Valuable Information Tool

The Government has to remove the constraints in the distribution of remotely sensed data so that farmers can access them easily and in time.

Agricultural production has remained nearly stagnant during the past 6–7 years, in spite of new technologies. It is also stated that land degradation has affected a large area perhaps due to intensification of agricultural practices, overuse of fertilizers, poor drainage in highly irrigated areas and soil salinity.

In spite of the projected surplus of foodgrains if productivity does not increase in the next 4–5 years, we may have to import foodgrains to feed the increasing population.

When the green revolution, during the mid-Sixties, paved the way for self-sufficiency by increasing productivity. A few agricultural scientists had warned us about the after effects such as land degradation that might result after some years. They said that we should develop alternative farming systems based on organic inputs to increase productivity without adversely affecting the health of the soil, to reach a level of reasonable sustainability from the point of food security. However, such efforts have been rather slow.

We have to monitor natural resources, particularly those affecting agricultural practices, combine traditional with modern methods and approach the problem in an integrated manner. The problem is compounded by natural calamities, such as floods and drought, global warming, deteriorating groundwater quality and availability, and heavy top soil erosion. We have still not been able to restrict water losses due to high evaporation and run-off and increase groundwater recharge to match the water level with the rate of extraction. All these require a reliable information system and efficient management practices.

Satellite Data

Data obtained from space-borne remote sensing satellites can be used to obtain adequate and relevant information covering a larger area at frequent intervals. Techniques such as GIS also aid decision-making and better planning. Our land-holdings are rather small and the use of advanced technologies has to be commensurate with this aspect and be highly participative in its application.

Application aspects (Plate 33.1) and studies made by several agencies which specialise in remote sensing rarely reach the people who need them. In the mid-Seventies, we started using satellite data which had a ground resolution of around 80 m and a spectral resolution of four bands, mainly in the visible and near-infrared regions of the electromagnetic spectrum.

Frequency of data was 18 days initially or nine days with two similar satellites. This improved to 16 days and eight days with an improved spatial resolution of 30 m, later. These were found adequate to map on a scale of 1:50 000 which was nearly satisfying the needs of district level information.

Finer mapping became possible with the French satellite having a 20 m resolution, multi-spectral three-band sensor and a 10 m panchromatic black and white camera. Even with this, we could reach only an effective mapping scale of 1:25 000 with some complex processing. This satellite data also had provision to obtain stereo pairs, with which one could see terrains in three dimensions with reasonable accuracy. With the advent of powerful computer systems and GIS software, packages creating digital elevation model (DEM) or digital terrain model (DTM) became possible to study the terrains and natural resources with a better perspective.

Improved Accuracy

The recent Indian remote sensing satellite gives data in a panchromatic mode of 5.8 m ground resolution, a 24 m ground resolution in multi-spectral mode including a short wave infrared band, and a wide field sensor with a coverage of about 800 km swath and a resolution of 185 m.

This was a quantum jump in the quality and applicability of remote sensing data. The panchromatic data was capable of giving stereo pairs, to study in three dimensions, frequently.

Inspite of all these, as these data were optical in nature, these could not be seen through clouds and so microwave satellite data which can penetrate through clouds was chosen.

Micro-level Planning

Thus, there is a variety of remote sensed data improving our capability to study, assess, map, monitor and manage our natural resources effectively and, to a great extent, quantitatively. As our land-holdings are small we need to apply the concept of micro-level planning and adopt the concept of micro-watersheds for the benefit of small land-owners.

There has also been an advocacy of small satellites and satellites with more spectral channels, but implementing them may not be practical or economical. However, higher spatial resolution satellites both in panchromatic mode and multi-spectral mode are coming in a big way from private companies.

Before the end of this year US companies are expected to launch satellites having a resolution of 2 or 3 m. Already the Russians have launched their original reconnaissance satellite with 2 m resolution.

Before the turn of the century, satellites giving 1 m ground resolution with stereo capability are predicted from several private companies in the US Nearly 50 satellites were launched during the last 30–35 years, and in the coming 10–15 years an equal number are expected to be launched. What percentage of the total data that is beamed to us by such a large number of satellites is being used? Around 10 percent only. In spite of this, the economics of using satellite data for managing our natural resources is good.

In India, the Department of Space and the NRSA have already initiated an integrated mission for sustainable development covering a large number of districts mainly based on natural resources and its planned use (See Plates 33.1 and 33.2). Some of the action programmes given in these reports might have been more academic or theoretical as many land-owners, particularly the smaller land-owners, may or may not adopt it both on account of the slow economic returns or their own needs. The quality of the information developed must also be strictly checked.

No attempt has been made to describe the technicalities of remote sensing or GIS. The aim should be to ensure faster accessibility and a better distribution system for remotely sensed data products to take advantage of such high quality data for use not only in India, but also elsewhere. To map and identify small holdings and individual parcels of land, we will have to resort to higher resolution data as in the use of aerial photographs, and maps within 50 km of coastal areas are restricted and cannot easily be obtained by all.

While it is known that satellite data is bound to come with resolutions nearly matching aerial photographs, it may be wise to think of removing the restrictions imposed by the government and ease the availability of maps and aerial photographs, except in very rare cases where national security may be jeopardised. For example, aerial photographs are classified as secret even if they are 20 to 30 years old, and the reasons for this *en masse* restriction are not clear and are not commensurate with the technological advancement that has taken place. Modern satellite based geotechnologies or geoinformatics is rapidly improving and dependence entirely on aerial photographs or restricted maps may not be a real constraint.

At the international level, the use of digital data and its analysis using different kinds of GIS software packages have advanced considerably as could be seen from some of the papers presented as well as some of the demonstrations in the recent international congress of the International Society of Photogrammetry and Remote Sensing, held in Vienna. Our knowledge-base in this field is also certainly growing but in application, we are lagging behind.

34

Geoinformatics and Human Resources

This brief discusses what courses in geoinformatics are available and recommends human resource development in this area.

Geoinformatics, geotechnologies or geomatics, all those terms mean the same thing, except that different countries use different names for the same activities. In India, geoinformatics and geomatics are used. Anna University terms its four year B E degree programme as Geoinformatics, perhaps following Germany's example. Recently a National Society of Geomatics has been formed. The term geotechnology is popularly used in the US All these terms are used to denote various methods adopted for identification of various features and measurements on the surface of the Earth, and its resources, putting them in an orderly manner using computers and generating information needed for planning. This may include remote sensing from satellite platforms or aerial platforms, GIS and their use for integrating information available in different formats or forms. This information is analysed and processed using computers.

Global Positioning Systems (GPS) are used for geographically coordinating various points on the surface of the Earth using a satellite constellation system called "Navstar" belonging to the US defence department and now available for civilian use. Autocad provides maps from digitally stored data and extracts different features, via photogrammetric technology using aerial photographs flown at different altitudes for preparing maps in digital form or in analogue form and encompasses all related activities to provide appropriate information to planners and for design and implementing developmental projects.

While such activities were carried out manually using traditional ground visits which are laborious and time-consuming in the past, the technological advancement that have taken place rather rapidly during the past two or three decades makes survey operations easy, at less cost and time. Today, more work is possible and field visits are few, thanks to data obtained from satellites and aeroplanes. Therefore, a variety of data has become available and it requires quite a bit of understanding and experience, both through regular academic programmes as well as practical training, so that enough human resources are available to take maximum advantage of these technologies. Advance in technologies is faster than developing its applications and developing trained human resources. This article attempts to bring out certain aspects of academic advancements being made in India, particularly by the Anna University, the only university to have introduced a four year geoinformatics programme in its department of civil engineering, leading to the B E /B Tech degree.

At first there were apprehensions whether students graduating in this subject would get suitable positions. These fears have proved unfounded. The first batch of about 15 students who passed recently got employment almost immediately, at slightly higher salary levels than other students. This confirms that the chances of getting employment are quite bright. While there have been some universities engaged in conducting post-graduate programmes in remote sensing, including some limited exposure to GIS, a comprehensive course and understanding of most parts of geoinformatics or geomatics is now coming only from the four year degree programme of the Anna University, with further specialisation if required.

In some countries, under-graduate degree programmes in survey engineering have been in vogue for a long time now, but in India, this subject was given very low priority. This resulted in utter disinterest by students and staff in this subject. As the price of land goes up and per capita availability of land decreases due to population growth, it becomes essential or even necessary to provide more and more accurate plans and maps so that everyone knows the exact amount of land one possesses.

It has been generally stated that for every 10 000 population group there should be one professional surveyor available, whereas in India we are miles away from this figure. Hence, development of human resources in these subjects in larger numbers assumes considerable importance for the future.

The syllabus and subjects taught covers generally eight semesters with subjects such as; mathematics, physics, chemistry, engineering practice and practical computer and workshop practice in the first semester which enables students in the initial stages to understand more of the engineering and professional knowledge to come in the next semesters.

In the second semester, besides humanities and mathematics, an introduction to remote sensing is included along with basic electronics and instrumentation and strength of materials with practicals. In the third semester, subjects like engineering survey, advanced mathematics, introduction to photogrammetry, statistics, informatics and some engineering geology are taught. (See Plates 34.1 and 34.2)

Fourth semester students get a higher dose of engineering survey, numerical methods, a higher level of photogrammetry, a higher level of remote sensing, initiation into cartography and survey adjustments. The fifth semester includes advanced engineering survey, photogrammetry, survey adjustments, introduction to geodesy, introduction of geographical information systems and geodetic computations are taught with practical exercises.

In the sixth semester students learn digital image processing, electronic instruments for surveying, hydrology and water resources and urban and regional development along with one elective topic of the students' choice. In the seventh semester, introduction to cadastral surveying, satellite geodesy, environmental engineering and one or two electives of students' choice, are imparted. The last and eighth semester includes mostly electives of students choice, some management concepts, a project work and some discussions on group basis.

There are nine electives which include, microwave remote sensing, close range photogrammetry, advanced cartography, remote sensing applications to hydrology and water resources, ocean engineering and ocean resources, environmental science, urban planning, earth sciences and agriculture and forestry.

Going through the above subjects in this four year degree programme of the Anna University, supported by the Institute of Remote Sensing (originally conceived as an extension unit), the students coming out are reasonably well informed. The teachers have had adequate exposures through training fellowships offered by Germany and the Netherlands.

However, the number of students likely to pass out every year is grossly inadequate to meet the professional needs of our country. With the computer becoming an indispensable tool and becoming available at affordable prices, most industries as well as earth resources managers will need qualified personnel. The NRSA has taken up a project called integrated mission for sustainable development of about 150 to 200 districts, and without sufficient professionally qualified and somewhat experienced people the task cannot come to the quality level that is expected. (See Plate 34.3) Therefore, human resources will have to be developed fast by introducing detailed and comprehensive programmes even at full time undergraduate level in universities and IIT's to meet the demand.

There are institutions such as the Survey of India, the Indian Institute of Remote Sensing, the various state cadastral survey organisations imparting knowledge in these areas, but there are limitations to such programmes. The Survey of India trains its own staff. After a two year training the officers qualify in an examination conducted by the Institution of Surveyors, a professional society. Some of these officers later undergo specialised courses in geodesy, photogrammetry and other such subjects. But the services of these officers are used by the Survey of India for their work. The Survey of India's office in Hyderabad also offers training to a few sponsored candidates from other departments but no private individuals seem to have been entertained till very recently. Similarly the Indian Institute of Remote Sensing (IIRS) of NRSA offers programmes for sponsored candidates, but does not cover all aspects of geomatics. The cadastral survey departments in various states have still to catch up with the modern technology, and they need a complete overhaul to enable proper management of land. They will have to enter into an era of digital cadastre and of numerical cadastre, which can define individual holdings in unambiguous terms, describe its quality and potential, and reduce disputes.

More under-graduate programmes are essential, so that by the turn of the century, we will at least have a nuclear of 100 to 200 graduates coming out regularly.

Land availability is becoming limited in terms of the needs of the growing population, and it is essential that we build up capacities to utilise the existing land optimally. In addition, certain institutions like the Institute of Remote Sensing, Anna University have been generating reasonably good information systems through the execution of a number of specialised projects, but their findings and reports are not easily available nor well disseminated. This is perhaps the case with other institutions also. An integrated approach to developmental planning and management will be crucial to our survival and to realise the concept of sustainability.

35

Satellite Surveillance for Drought Conditions

Drought conditions can be monitored using data obtained from satellites. This system provides efficient and timely monitoring capability by integrating the time-effectiveness and objectivity of space observations with details of ground perceptions.

One of the natural calamities that affect us is lack of normal rainfall and consequent drought conditions which in turn affect agricultural productivity. We will now see how drought can be monitored using data obtained from satellite altitudes. Drought is a phenomenon that occurs rather slowly and predicting it is difficult. Lack of rain and severe dry conditions can lead to food shortage. Till about 1987, when we had to face severe drought, the NRSA initiated a programme of using satellite data to monitor vegetation cover of large areas, with respect to agricultural activities. This programme implemented with the support of the Ministry of Agriculture called National Agricultural Drought Assessment and Monitoring System (NADAMS) was carried out by NRSA's Water Resources division based on the NOAA (National Oceanic and Atmospheric Administration) series of satellites launched by the US. The data was obtained from the sensor called advanced very high resolution radiometric data (AVHRR). The data is received everyday.

A single satellite of the polar orbiting TIROSN series (television infrared very high resolution radiometer) produces pictures of a large area and gives a complete coverage of the entire surface of the Earth daily. A geostationary weather satellite gives images as often as every half-an-hour, of a fixed portion (30 to 40 percent) of the surface of the Earth. One example is the INSAT series of satellites from where we get pictures in newspapers and also on our television every day.

The TIROS-N. ABVHRR Satellite has four channels, i.e. visible range (0.55–0.9), near infrared (0.725–1.0) middle infrared (3.55–3.93) and thermal infrared (10.5–11.5) and thermal IR (11.50–12.50) micrometer wavelength. The spatial resolution is 1.1 km and has a swath of about 3 000 km. The AVHRR data, although primarily meant for meteorological purposes, can be used to know the vegetation and its vigour, water and moisture conditions, etc. This is used for monitoring the drought conditions that exist at a particular time. The information is passed on to the authorities concerned. Drought management to be effective, requires reliable and timely assessment of conditions.

Previous drought conditions were characterised by ground observations of rainfall, aridity and agricultural conditions in terms of cropped area and yield. "Rainfall as an agricultural drought indicator, is limited by the sparse ground observations as well as the lack of spatially and temporally unique relationship between incident rainfall and vegetation development." Though the ground observations by the state departments of Agriculture and Revenue of agricultural conditions are exhaustive, such a system involves a significant amount of subjective judgment at various stages. Therefore, the NADAMS will provide more efficient and timely monitoring capability by integrating the time effectiveness and objectivity of space observations with the details of ground perceptions, say NRSA scientists.

Because it covers the entire country, NADAMS envisages the use of data from the NOAA satellites AVHRR sensor with 1.1 km resolution at nadir and having every day revisit capability. It has also been possible to use higher resolution satellite as supplementary information. Since there will be cloud cover in some parts of the country during the agricultural season, NADAMS uses NOAA -AVHRR data generated on a period of time. This generally gives an indication of the greenness of an area and approximate biomass. Ground data on rainfall and other agricultural parameters such as the crop area is also used in the operational use of NADAMS.

The method was validated during the 1989 cropping season. Bi-weekly drought bulletins during the June-December, 1989 were generated covering about 246 districts in 10 states, but only six states were covered in June-October.

These bulletins contained the vegetation index map of the state, along with a brief report. It also covered a comparison district-wise of the intensity of greenness during this period and corresponding previous years.

The bulletins gave an assessment of relative severity levels. The NRSA reported that very encouraging feedback was received from many users of such information particularly the district level authorities. (See Plate 35.1)

Better meteorological rainfall estimation, and improved rainfall forecast could significantly enhance drought monitoring capabilities as well as its short-term management. Soil moisture monitoring is yet another aspect that is yet to take off at an operational level. The NRSA and Space Application Centre (ISRO) are actively engaged in using microwave remote sensing data obtainable from European remote sensing satellite at an experimental mode to see to what extent soil moisture could be estimated and used in the NADAMS project, to make it more effective.

During drought, physiognomic changes vegetation may become apparent. Satellite sensors are capable of discerning many such changes through spectral radiance measures and manipulation of such measures into vegetation indices, which are sensitive to the rate of plant growth as well as to the amount of growth (Curran, 1980). Such indices are also sensitive to the changes in vegetation affected by moisture stress. The visible and near infrared bands on the satellite multi-spectral sensors allow monitoring of the greenness of vegetation. This property is used in the case of monitoring drought, as the stressed vegetation and other bare ground, water, etc. reflect differently. Besides, moisture stress in vegetation, resulting from prolonged rainfall deficiency, is reflected by lowering of vegetation index values. Such decrease could also be caused by other stresses such as pest/disease attack, nutrient deficiency or soil geo-chemical effects. But this does not show up well in coarse resolution data which covers very large areas at a time. (See Plate 35.2)

I may mention here, that the foreign crop condition division of the US department of Agriculture, the Canadian Wheat Board, the statistics Canada crop condition assessment project, and global information and early warning system (GIEWS) of the Food and Agricultural Organisation of the United Nations (FAO-UN) are using the vegetation index data as one of the major inputs for assessment of agricultural condition in different parts of the globe. The NADAMS project of the NRSA also considers two weekly vegetation index data as an important input along with rainfall data and aridity indices for countrywide assessment.

Reliable drought interpretation requires a geographical information system (GIS) approach, since the topography, soil type, spatial rainfall variability, crop type and variety, irrigation support and management practices are all relevant parameters, as stated by senior scientists of the water resources division of NRSA, who are directly involved in this project. A pilot study is being undertaken in this direction.

In conclusion, the Indian Remote Sensing Satellite (IRS-1C) sensor called a wide-field sensor (WIFS) data available since January 1996 (This data has a swath of around 800 km and spatial resolution of around 185 m) and from the IRS-P3 from about March 1996 are providing repetitive coverage of large areas every five days, which provides an impression of the vegetation dynamics and when combined with the NOAA satellite data, is bound to enhance drought monitoring capabilities. It is hoped that over a period of time, it may become possible to forecast well in advance what the drought conditions are likely to be, to enable decision-makers to take appropriate steps in the right direction from the point of view of supporting a sustainable food security system.

36

Crop Forecast from Satellite Data

With technological advances, it is now possible to estimate the total area under cotton crop using satellite remote sensed data. These modern tools have made it possible to improve productivity both in terms of quality and quantity.

One of the important commercial crops in India is cotton and it is used mainly by the textile industry. Byproducts like oilcake, hulls and linter have appreciable economic value. India's cotton cultivated area is estimated to be one-fourth of the world's area under cotton crop. However, the average yield is only around 287 kg/ha against the world average of 567 kg/ha.

Cotton is grown in Punjab, Haryana, Rajasthan, Gujarat, Maharashtra, Madhya Pradesh, Karnataka, Andhra Pradesh and Tamil Nadu. Maharashtra occupies nearly one-fourth of the entire cotton area, but has the lowest yield of around 100 kg/ha, whereas Punjab has the highest yield of around 500 kg/ha. It is also reported that the prospects of a substantial rise in the output of cotton during 1996–97 appears to be bleak, in spite of all efforts.

There are several agencies which estimate cotton production in India such as the Ministry of Agriculture, the East India Cotton Association and the Cotton Corporation of India. Information is compiled by the Cotton Advisory Board. However, it has been found that the estimates vary considerably, making it difficult for decisions to be taken on import/export by the Ministry of Textiles.

An attempt is made in this article to estimate the total area under cotton crop, using remote sensed data and its analysis by various agencies both at Central level like Space Application Agency, ISRO, and some of the State Remote Sensing Application Centres such as Maharashtra, Punjab and Haryana, in arriving at a more reliable figure.

The assessement of the area under different agricultural crops, such as wheat and rice has been encouraging using remote sensing techniques and some more work is going on to predict precise production levels or yield estimates, well before the harvest which has high commercial implications, and also to ensure food security. These might shortly lead us to operationalising the techniques of advanced prediction of food crops towards which various organisations are working in India, using better satellite data, like the IRS-1C.

Sometime in 1991 or so, a project on cotton acreage and condition assessment using remotely sensed data was taken up by the Department of Space (SAC-ISRO) at the request of the Ministry of Textiles with the following objectives.

a) To estimate pre-harvest acreage under cotton crop in the major cotton growing districts of Andhra Pradesh, Gujarat, Haryana, Madhya Pradesh, Maharashtra and Punjab.

b) Monitoring the condition of cotton crop on a monthly basis.

c) To initiate yield model development by creating database of vegetation index profiles and other yield-affecting variables.

The study used information on drainage, climate, land use, etc, all by using satellite remotely sensing data, then available, viz., the IRS-IA, supported by appropriate ground truth measures. The study was carried out jointly with Maharashtra State Remote Sensing Centre at Nagpur. A similar study was undertaken in Punjab in four districts, using IRS-1C data and digital techniques.

Discriminating cotton crop from other vegetation did not pose any serious problem in using satellite remote sensed data. It was also found that the cotton estimation in terms of area was found to be quite accurate when compared to the conventional techniques which takes a longer time for acreage estimation. This was done jointly by the Punjab State Remote Sensing Application Centre and by the Space Application Centre of ISRO. Similar work was also carried out in other places, and in all cases it was found that the use of remote sensed data not only speeded up the work but also provided a much more accurate estimate in terms of the acreage under cotton crop (See Plate 36.1 (a) and (b)).

In the beginning of 1995, the Ministry of Textiles requested the Department of Space, Government of India, for production estimates for the entire country based on acreage and with a yield model developed by SAC-ISRO. The Ministry of Agriculture requested the Department of Space (DoS) for the inclusion of 43 districts of Intensive Cotton Development Programme (ICDP) under the CACA project (Cotton Acreage and Condition Assessment). In actual fact, the study finally included 53 districts instead of 43 districts initially suggested.

The objectives were: i) Pre-harvest acreage estimation for all project districts. ii) Condition assessment and yield model development using single/ multiple ratio/regression methods.

While some limitations were encountered, with respect to obtaining collateral data to be included in the yield model, the procedure adopted provided yet another opportunity to prove the strength and power of satellite remote sensing data in the area of cotton crop, which will ultimately lead to considerable economic benefits. With improved and better satellite data availability, such as IRS-IC and improved procedures, it is almost certain that this technique will be operationalised very soon and the technology passed on to the State government organisations using remote sensing for their economic benefit and policy formulations.

While the statements have proved the ability and usefulness of finding acreage under cotton crop, a more powerful use to support a variety of decisions, based on integrating several factors, called the GIS has emerged during the past decade and a half, which if used carefully can lead us to adopt measures to increase yield to a great extent.

One of the suggestions that is being made by some very eminent agricultural and soil scientists to increase productivity is use of crop rotation whether it is under irrigated or rainfed areas. It is also felt that while use of nitrogen fertilizer is a general practice, it may be used in excess and thus it may be worthwhile using about 50 percent organic fertilizers, which can also activate better yield.

The practice of integrated pest management, integrated chemical control and the use of biological-based pesticides may also give scope to higher productivity. Likewise, there are several other factors that could be identified which if used after considering various alternatives through use of a computer and with a well defined GIS software, is likely to throw open several simulation models to enable some practical approach. These modern and technological tools have now become available and are bound to improve productivity, both in terms of quality and quantity. These statements have been made as a measure of furthering research and development to enable India become one of the leading cotton producers.

37

Sentinel in the Sky

The value of biodiversity lies in its contribution to the ecological systems which maintain life and economic activity. The destruction of habitats rich in biological variety also disrupts hydrological cycles and leads to hardship. This brief discusses the role of remote sensing in biodiversity management.

Biodiversity is at a greater risk now than it ever was. The continuing rapid rate of habitat loss and degradation have led to unprecedented threats to biodiversity on which the livelihood of local communities and national economies ultimately depend. The value of biodiversity lies in its contribution to the ecological systems which maintain life and human economic activity; in the utilisation of components of this diversity by humankind for food, medicines and raw materials; and in its cultural, spiritual and aesthetic significance for human societies on a sustainable basis. The destruction of habitats rich in biodiversity also disrupts hydrological cycles and leads to hardship. India has a variety of geological and bio-geographical zones, viz. trans-Himalayan, Himalayan, the Indian desert, semi-arid tracts, the Western Ghats, the Deccan peninsula, the Gangetic plains, Northern terrain, islands and coasts.

All along the coasts, mangrove wetlands, a major natural resource for the ecological and livelihood security of the coastal communities, have been severely degraded. Similarly, in the inland areas including hilly terrain, soil is being degraded for the cultivation of high yielding crop varieties which require high inorganic fertilizers. In biologically rich areas, such as tropical rain and moist forests, genetic resources are being depleted at a fast rate due to human induced stresses such as deforestation and their developmental activities.

To study and evolve remedial measures to the extent possible, the M S Swaminathan Research Foundation has been conducting a variety of research and demonstration projects, where remotely sensed data is used for planning, implementation and monitoring activities. A case-study related to the conservation and management of mangrove wetlands of Pichavaram in which remotely sensed data was effectively utilised is given below. The Pichavaram mangrove wetlands are located at the northern end of the Cauvery delta, in Tamil Nadu on the south-east coast. Compared to other mangroves, these mangroves are rich in species and diversity, especially of true mangrove plant species. Fourteen exclusive species have been recorded, out of which conservation of three species namely, *Rhizophora apiculata, Rhizophora mucronata* and *Rhizhopora lamarackii* require urgent attention. These species are no longer found in localities like Muthupet and have become endangered in other places, such as the Sunderbans (West Bengal and Bangladesh), the Godavari and the Krishna mangroves. Realising these facts, government agencies have been taking a number of steps in the last 10 to 12 years to manage these mangroves on a sustainable basis. However, many of their attempts have not succeeded substantially.

Remotely sensed data plays a crucial role in solving these problems, as it is available periodically. The assessment of the degree of degradation of the Pichavaram mangroves has been possible to a large extent using remotely sensed data.

The Pichavaram mangrove forests were declared a reserve forest in 1897, and during that time the total area was about 700 ha. However, for various reasons, the extent has reduced substantially. Though the Government agencies were aware of the degradation, they were unable to quantify the area under the degraded category of mangroves since conventional methods of surveying were not possible in the mangrove wetlands. This lack of information, along with other reasons, hindered the Government agencies in preparing a reliable conservation and management plan. However, using Indian Remote Sensing Satellite (IRS) data, it was found that nearly 62.8 percent of the Pichavaram forests were degraded between 1897 and 1994 and the rate of degradation was about 5 ha a year. This information was passed on to Government agencies which made them realise the urgency for the restoration and conservation of this important mangrove.

Analysis of the causes for the degradation of the Pichavaram mangroves using remote sensing revealed several reasons.

The government agencies and research institutions working in the area found that cattle grazing and illegal felling were the main causes for degradation. However, remote sensing analysis revealed that the middle portion in the area was most affected. This indicated that apart from grazing and felling, there could also be some other reasons for such degradation since these areas are not easily accessible. Further, ecological studies revealed that changes in topography into some sort of "trough" shaped resulted in the stagnation of tidal water which reduced the growth of the mangroves in these areas.

The Pichavaram mangroves were in the past protected from wave action by a long stretch of sandy beach. However, a comparison of the Survey of India toposheet and the remote sensing imagery of IRS-1B LISS II shows that the breadth of this beach was reduced by 550 m, between 1930 and 1970 and about 150 m between 1970 and 1992. The rate of erosion is estimated at about 13 m a year. This indicates that there is a danger of the Pichavaram mangroves being exposed directly to the open sea when the wave action is very high and which will hamper the regeneration of mangrove plants. Apart from these, a study of topographical map sheets and recent remotely sensed imagery also indicate that in these areas, erosion and sedimentation occurs almost simultaneously.

Considering these aspects, a long-term management plan is necessary. It is being attempted in the Pichavaram area where both remotely sensed data and GIS are used to advantage. With such initial experience, a more detailed and long-term study of conservation and regeneration of mangroves is being attempted by the M S Swaminathan Research Foundation. It is to cover areas in Tamil Nadu, Andhra Pradesh and the Orissa coast (See Plate 37.1). A well-equipped facility has been set up for this.

In this context, it would be desirable to undertake an integrated coastal environment study to evolve methodologies to arrive at suitable management practices using periodic remote sensed data.

This would enhance the understanding and capacity for national action in the conservation and sustainable management of coastal wetlands.

The recently launched satellite, IRS-1C (December 28, 1996), carrying three different sensors offers unique application opportunities and provides an almost near real-time information system for regular monitoring activities of various natural resources.

38

Remote Sensing by RadarSat

Data from the RadarSat is useful in crop assessment studies, land use monitoring, base mapping, oil spill detection in the oceans, aquaculture site detection, geology and hydrology to name a few areas. This article focusses on the aspects of application of dependable information from a Canadian product.

Since optical satellites cannot see through clouds and in conditions of poor visibility, it was soon found necessary to have satellite remote sensed data for the study and analysis of floods and flood-prone areas, limited soil moisture assessment and regular agricultural monitoring, particularly during cloudy conditions.

Microwave sensors or cameras use their own energy systems, unlike optical sensors which use the sun's reflected energy through the visible regions of the electromagnetic spectrum. In fact, this was one of the shortcomings of optical remote sensing where it is not possible to obtain pictures when an area is covered by cloud and when data is needed.

The first microwave satellite data became available when the European Remote Sensing Satellite (ERS-I) went into orbit in 1991. This was followed by ERS-II in 1995, which carried more or less similar sensors as ERS-I and had an additional sensor as well.

In November 1995, Canada launched the RadarSat with a view to providing a unique new source of microwave data for a variety of users. The use of the synthetic aperture radar revolutionised microwave remote sensing and made possible high resolution imaging from space, irrespective of weather conditions, of our Earth's resources for specific applications. The basic constraint of a real aperture radar was overcome by using the concept of a synthetic aperture in obtaining a long aperture, synthetically using a small physical aperture. By using a synthetically processed long aperture, a much higher spatial resolution constantly, even from space, was obtained.

Historically, the use of the RadarSat dates back to the late Seventies when Canada used it to monitor it's frontier regions, with government-industry cooperation. Indo-Canadian cooperation in the Indian remote sensing industry is more oriented towards agriculture, environmental and other mapping needs, as this tool makes it excellent during the rabi and kharif monsoon seasons, to be a complementary source of information to Indian remote sensing data. It has been reported that the combined use of RadarSat synthetic aperture data and optical IRS data can improve crop acreage estimation and yield forecasting. Another important area in which SAR data is very useful is in flood mapping, and flood monitoring.

"RadarSat is an advanced Earth observation satellite developed by the Canadian Government and is Canada's first remote sensing satellite. Its data products will be accessible to commercial, government and scientific users. Its life-span is scheduled for five years. Equipped with a Synthetic Aperture Radar (SAR) system, data will be transmitted from the satellite in real-time to a ground station or stored on RadarSat's tape recorders for subsequent downlink. RadarSat's SAR provides significant advantages in viewing the Earth under conditions that preclude observation by aircraft or optical satellite sensors." This satellite orbits at about 800 km from the Earth in a near polar (98.6) Sun-synchronous orbit. This was how it was introduced.

Raw RadarSat data consists of unprocessed, complex SAR data which will require extensive processing before images can be obtained. Geo-referenced products are geometrically corrected to compensate for the Earth's curvature. Latitude and longitude information for each line of data is also included in the digital information on tape.

Geo-coded products supply precise positional information which is corrected to conform to one of 30 standard map projections (e.g. UTM, polar, stereographic and Lambert conformal). A map projection is a system of drawing horizontal and vertical lines in such a way on paper to represent a curved surface and it is different in different locations and countries mapping systems. These products are available in digital form also.

RadarSat also claims revisit capability of less than 5 to 10 days, depending on resolutions such as 10 m and 30 m respectively. Its utility in agriculture in crop assessment studies, land use monitoring, soil condition monitoring in terms of temporal changes, soil moisture, etc., are unique and valuable. In cartography it can be used for base mapping and also for terrain evaluation studies in terms of 'x', 'y' and 'z' coordinates. It has also been found very useful in coastal zone monitoring of coastal changes, oil-spill detection in oceans, aquaculture site detection, bathymetric studies, etc. In forest management studies, it is also a very useful information provider. Other areas of its uses include geology, hydrology and ocean parameters.

In the area of disaster assessment and accurate monitoring such as earth quakes, volcanic eruptions, oil-spills and flood events, the use of RadarSat claims to be unique as it has the ability to provide imagery within a few days of its occurrence. For strategic and tactical planning, RadarSat provides very valuable information.

In India, the Space Application Centre (SAC) of the ISRO and the NRSA have carried out extensive experimental studies using microwave data in most of the resource areas as pointed out earlier, and such studies are being further extended combined with data obtained from comparable optical sensors. The use of ERS Synthetic Aperture Radar (SAR) data for the identification of paddy, sugarcane, cotton and groundnut crop in four different test areas representing various agroclimatic and crop distribution characteristics by SAC (ISRO) on 1993 or so, has shown promising results, particularly when data obtained on two different dates was used.

Thus, it can be seen that RadarSat data, being available in a variety of more resolutions and time intervals, is expected to play a useful role in the area of use of remote sensing especially agriculture and related areas, disaster assessment and forestry related subjects. Its use, combined with other types of data obtained from optical sensors, and integrated with other ancillary data, is bound to play a useful role in the management of the Earth's resources, specially when it is used in a GIS environment.

It is also claimed that in the area of cartography and topographical map making, a single data source can offer customised planimetric and altimetric information for use in creating and updating existing topographical maps. To visualise data of the topography in three-dimensional perspective, Digital Elevation Model (DEM's) on the computer can be derived from RadarSat data either through the technique of radargrammetry or interferometry.

Monitoring land use change using RadarSat data has been remarkable. Multi-temporal black and white data can be processed to give a coloured picture output. These were made available by the Land use and Cartography Group of NRSA, Hyderabad (See Plate 38.1). Another illustration included in this article was provided by the NRSA, Agriculture and Soil Group, pertaining to the Rajamundry area in Andhra Pradesh (See Plate 38.2).

The coastal zone is ideally suited to Synthetic Aperture Radar (SAR) satellite Earth observation techniques because of the synoptic coverage provided and the ability of SAR to provide reliable monitoring independent of light and weather conditions. It is being used extensively in ship surveillance, oil-slick detection and ocean features detection.

It is claimed that the use of Radarsat data for hydrological studies, identifying geological features, developing a DEM for crop monitoring particularly rice crop, environmental monitoring and disaster assessment can give good results. However, work is still being carried out by some of the leading organisations in India in these areas.

In conclusion, it can be stated that the Synthetic Aperture Radar data has been able to provide a versatile data system of the Earth's resources both by studying it singly or in combination with other complementary data, to derive more authentic and dependable information and to continuously monitor the same.

39

Remote Sensing to Help Boost Tourism

This article suggests that remote sensing can be used to promote tourism by identifying places of interest and minimizing environmental damage, thereby leading to a lucrative Tourism Industry.

People the world over nowadays just do not take holiday trip for rest and recreation. They like to educate and enlighten themselves. So, all countries try to tap their tourism potential not only to entertain visitors, but also to highlight their culture and heritage in their natural surroundings, at the same time ensuring that the environment is not seriously affected by man-made activities.

India is a country with a very rich, very old and highly diverse cultural heritage, offering ample opportunity to develop sites to attract tourists. There is abundant scope to develop newer areas which can satisfy people who look for innovative sites to visit. India exhibits a vast diversity in its geographical layout, physical characteristics and cultural practices. With a population of almost a billion, this diversity is witnessed in the social structure, essentially in the dialect, costumes and clothes, food habits, and religious beliefs. Alongside, the country boasts of a rich heritage and ancient civilisation that has nurtured economic, social and cultural development.

These aspects tend to capture the interests of people worldwide who look to an enriching, yet inexpensive, vacation. Tourism as an industry in India is still in its early stages. Travellers from far and wide may have made their way into different parts of India, yet it has not been until the last decade or so that tourism has been focussed on as a potential source of revenue. The more money invested in it, the more revenue will it earn, especially foreign exchange, besides enlarging the scope of employment opportunities.

In the development of this industry certain important factors need to be considered. Primarily, even a naturally attractive place or a historically eventful well connected site can be marketed as a potential product. In developing such sites, we may have to develop infrastructure, reduce travel barriers, particularly for foreigners, and attract target groups by careful study of their needs. Tourist markets may either be domestic or foreign and for both, the scope in our country is enormous, provided we select potential sites from all angles to ensure stability and proper environmental protection and prevent adverse impact.

Tourism could be of different kinds, like

a) Heritage tourism

b) Adventure tourism

c) Historical tourism, including archaeological sites and old battle sites.

d) Cultural tourism

e) Sports tourism

f) Eco-tourism and

g) Architectural and heritage tourism.

While a variety of activities needs to be undertaken to develop and promote tourism, we have to consider the environmental impact that may result in such activities, besides carrying out detailed studies to identify suitable new spots. We have to map and monitor changes that may take place in those areas and develop an effective management plan to ensure that the environmental degradation aspect is taken care of.

In the last two decades, space technology has come up with remote sensing from satellite platforms. This has become a potential and powerful tool to observe and study any area on the surface of the Earth. In one picture we can see vast areas and with a potential to identify details as small as 5 or 6 m. Since such data is received from satellite altitudes in digital form, we can either analyse this data using a computer or convert the figures in to colourful pictures so that they can be seen.

Fortunately, India is one of the few countries where remote sensing technology has been used to great advantage.

As an example of effective use of satellite remote sensing technology, I give a brief account of what the land use and cartography group scientists of the NRSA did covering an area called Patnitop in the Kashmir Valley in 1994, with a view to opening up tourism as requested by the Jammu and Kashmir government, and in particular, the Chief Executive Officer of the Patnitop Development Authority. A study of the physical factors of the area was easily done by using remote sensing techniques, essentially a typical object -oriented land use/land cover studies. They can reveal changes that are likely to occur when tourism gets a boost, leading to a comprehensive management plan, incorporating other relevant attributes (See Plate 39.1).

The Kashmir Valley is well known for its natural beauty which attracts tourists from all parts of the world. It is but natural that the Government of Jammu and Kashmir is trying to develop newer areas. It is in this context that the NRSA undertook a detailed land use/land cover study of the Patnitop area, which could be used to prepare a master plan to develop a viable tourist spot with the least environmental side effects.

"The main objectives of the study area are:

i) To prepare land use/land cover maps on a scale of 1:25 000, using the French SPOT multi-spectral satellite data

ii) To estimate the extent of area under different land use/land cover categories using satellite imagery

iii) To prepare a thematic map volume and report on the observations and recommendation for land use/land cover management" (NRSA Project Report).

Besides using appropriate satellite imagery, the corresponding Survey of India topographical maps were also used as supplementary positional data. In addition, the interpreted details were subjected to adequate ground truth to ensure better reliability of the final map. The procedures adopted were simple visual interpretation techniques, similar to what is done in all other such studies.

Ultimately, in order to get an integrated picture, all the maps and other collateral data were entered into a GIS, so that it could support specific decisions to be taken by the authorities in the entire process of development and to know what should be done and what should not be done at different stages. Suggestions were also given in the report for consideration by the development authority.

The report was received with great enthusiasm by the development authority as a comprehensive study could not have been made at such a short time, without the use of remote sensing and GIS.

While remote sensing has proved itself as a versatile tool in mapping, monitoring and managing natural resources, combined with other attribute information and using GIS techniques, an integrated information system has become more valuable in all development activities, including developing newer areas for tourism, as travel and tourism is an ever-expanding activity for the human race, besides being a lucrative, commercial activity also.

40

Education in Space Science

With the aim of sharing the benefits of space technology with developing countries, the U N decided to establish regional centres around the world. One was established in India due to the advancements the country had made in all aspects of space sciences and remote sensing. This brief reveals how it happened.

The tangible and intangible benefits of space technology observed during the past two or three decades, in understanding various natural environmental processes, prompted the United Nations to establish regional level centres, specifically for the benefit of developing countries. The first of such centres, named the Centre for Space Science and Technology education in Asia and the Pacific (CSSTE-AP) was established in India at Dehra Dun for the Asia-Pacific region by the end of 1995, with the support of the Department of Space.

In this context, it is relevant to give a brief background about the Institute of Remote Sensing that was first started as the Indian Photo Interpretation Institute (IPI) as a constituent unit of the 250 year old Survey of India, in 1996, with help from the Government of Netherlands. This was through its International Institute for Aerospace Survey and Earth Sciences, popularly known as the International Training Centre (ITC), specially meant to cater to the developing countries, in the area of study of natural resources, and also topographical mapping (Photogrammetry) using aerial photographs, sometime in 1951, as space-sensed data had not started arriving at that time. The first director of ITC Netherlands was also once the Prime Minister of Netherlands, and as far as I know he preferred to be Director of ITC, than being the Prime Minister.

In 1963 or '64, there was a conference at Tolouse in France, where emphasis was placed on the use of aerial photographs as a medium to be used for surveying various natural resources, besides being used for topographical mapping. These officers took the lead in discussing with the senior professors and the Dean of the ITC Netherlands a proposal to establish a similar institute called the IPI at Dehra Dun, for which the Government of India, and the Government of Netherlands readily agreed upon. Consequently the IPI came into being in 1966 at Dehra Dun, the headquarters of the Survey of India.

This institute has been imparting free training to a variety of resources scientists sponsored by various universities, government departments and other recognised organisations. The trainees however have to bear their living expenses. The main divisions are, Photogrammetry (as a basic need), Geology, Forestry and Soils, with an intake of about 60 trainees each year, and a diploma granted. In the initial stages the best trainee in each division, was sent for advanced training to the ITC at Netherlands, under a scholarship. The main beneficiaries were the Geological Survey of India and State Geological Departments, the central and state soil scientists belonging to agricultural departments, some of the universities and other institutions, mainly belonging to the public sector or government.

A hostel is also available on the campus and some foreign trainees are admitted from some developing countries, sponsored by international organisation of the respective governments or under a Government of India scholarship. There is no bar on private candidates, but they have to pay fees and fulfill the minimum standards needed for admission.

This institute became popular, not only in India, but also in other parts of the world and the Dutch Government through the ITC has continued to maintain close ties.

Initially, the staff came from various government departments after training at ITC, Netherlands and for the first four or five years, its experts were deputed on a long term basis, till the curriculum, etc. stabilised. I was posted to this institute as Head of Photogrammetry for three years in the early Seventies, besides being the deputy director. It grew from strength to strength as it was found at that time, that the natural resource surveys carried out using aerial photographs, proved to be at least five or six times faster and cheaper, although ground visits were required for accuracy (See Plate 40.1).

This technique to a great extent eliminated subjectivity in the final survey and assessment and has now become an accepted technique all over the country, except where it becomes difficult to obtain aerial photographs, due to its inherent classified nature by the Government of India.

During 1990 or so, with the rapid growth of space technology particularly in remote sensing and GIS, the United Nations decided to provide training facilities for developing countries where the use of these techniques was economical and an urgent necessity.

After surveying several parts of the world and in view of the rapid advances that India had made and having a very modern institute for training, the UN decided to establish a Centre for Space Science and Technology Education in Asia and the Pacific with support from the Department of Space, Government of India, at Dehra Dun, in view of the existing Indian Institute of Remote Sensing (IIRS), starting the first post graduate course in Remote Sensing and GIS from April to December 1996, with a set of standard curricula to include all aspects of space sciences.

The training programme is not only hosted by IIRS at Dehra Dun, but also at the Space Application Centre at Ahmedabad (SAC) of the ISRO and the Physical Research Laboratory (PRL), Ahmedabad, as the nodal institutions in different aspects of space sciences, such as satellite meteorology and global climate, satellite communications, space science, etc., Additional courses have been going on since then, attended by several developing countries in the region and improvements taking place regularly (See Plate 40.2).

The institute has a governing board, presided by the Secretary of the Department of Space, and officials drawn from Korea, Indonesia, Kazakhstan, Kyrghystan, Malaysia, Mongolia, Nauru, Nepal, Republic of Korea, Sri Lanka, Uzbekistan, besides the UN. The centre is now present headed by Prof B L Deekshatulu, erstwhile director of the NRSA, and the institution is known the world over. The next step perhaps would be to either elevate the IIRS and the CSSTE-AP to the level of a deemed university, or attach them to some of our progressive universities to enable awarding post-graduate degrees. In fact, the IIRS is recognised by a few universities as a centre of excellence for carrying out research towards the doctoral degree.

41

Planning out Infrastructure

If we look at the developed countries, we find that their economic prosperity has depended on an efficient infrastructural network. India too, being a large country, needs such a system. This brief dwells on the function of remote sensing in facilitating investigations, a prerequisite for efficient and cost-effective design and construction.

The Oxford Encyclopaedic English Dictionary defines infrastructure as "the basic structural foundations of a society or enterprise, a substructure or foundations such as roads, bridges, sewers, etc., regarded as a country's economic foundation" and "permanent installation as a basis for military, and other operations." We can add to this railroads and power lines and other such essential communication systems which enable better living conditions at a better economic and social level.

Here, we shall deal with how the country's improved and accelerated economic development and expanded globalisation concept necessitates the building of super highways and the need for essential investigations to choose the optimum route which is a prerequisite for design and construction.

If we look at the developed countries, we find that their economic prosperity has, over the years, depended on their efficient infrastructural network, both on the surface as well as aerial. India being a large country, needs an efficient and well maintained rapid transport network in the form of wide roads, railway systems as well as inland and coastal waterways besides an aerial transportation network.

The government's policy is to build wider and better surfaced roads in the shortest possible time, particularly connecting major cities. For this, the Government may perhaps seek expertise and assistance from other countries and monetary help from international agencies. In certain cases, acquiring land may pose problems, as it will entail the displacement of people, particularly those in rural areas. The need to further build feeder roads from such large highways to connect interior areas may also be needed over a period of time.

Initial investigation will include a study of the land that is covered by a suitable corridor, so that further investigations and surveys could follow, before a design is made and detailed estimates are arrived at based on up-to-date information.

It is here that remotely sensed data becomes relevant, with its inherent advantages in providing an almost near real-time scenario of the area/route under investigation. In some cases, aerial photography and land-based surveys may become necessary, but with improved satellite data, these more expensive methods can be considerably reduced and rapid decisions taken.

The Indian Society of Remote Sensing conducted a national symposium with emphasis on infrastructural development during its annual convention in November 1997. It highlighted what can be achieved in the infrastructural area, with particular reference to the latest IRS-1C and -1D data, which provides the world's best civilian spatial resolution data in panchromatic mode of 5.6 m and multi-spectral colour pictures at a resolution of about 24 m resolution, and other types of data. While the data structure of IRS-1C and -1D is acknowledged as the best civilian satellite data for a variety of natural resources survey and management besides periodical monitoring, its usage for a more detailed study came from the 5.8 m spatial high resolution data, with the capability of stereo viewing where necessary.

In the past, the French SPOT Remote Sensing Satellite did provide similar panchromatic data on a 10 m spatial resolution, but the IRS-1C panchromatic data proved to be superior.

At the symposium at Hyderabad, several examples of successful work relating to infrastructure came up. Road work, town planning activities, pipeline and similar work were quoted, confirming the real worth in economic terms of using such data.

The lead paper by a private entrepreneur, belonging to the recently established Indian Resources Information and Management Technological Limited at Hyderabad brought out how in the coming years, remote sensing data will be an indispensable tool as the market share of infrastructural activity may reach 35 to 40 percent of investments, in view of rapid industrialisation, agricultural development and opening up of the economy.

Some of the successfully implemented projects include:

A feasibility study to identify a suitable area for a new township in the Bangalore metropolitan region, using remotely sensed data combined with other relevant information.

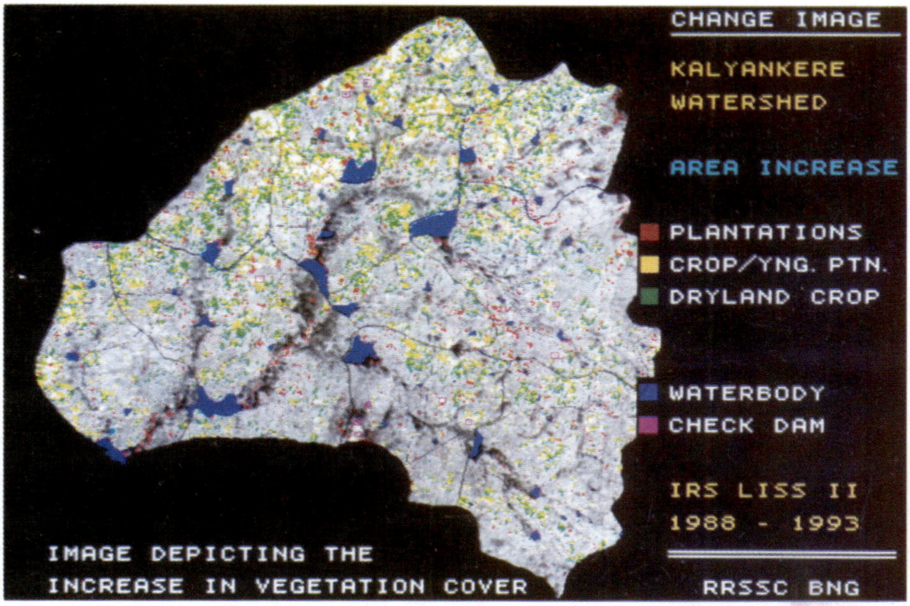

Plate 29.2 Image depicting the increase in vegetation cover.

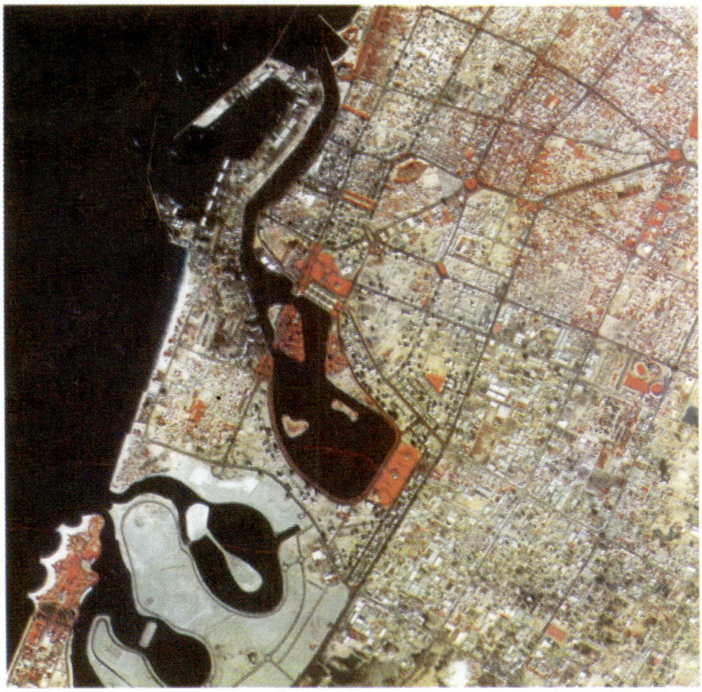

Plate 30.1 Image of Sharjah city; shows various features.

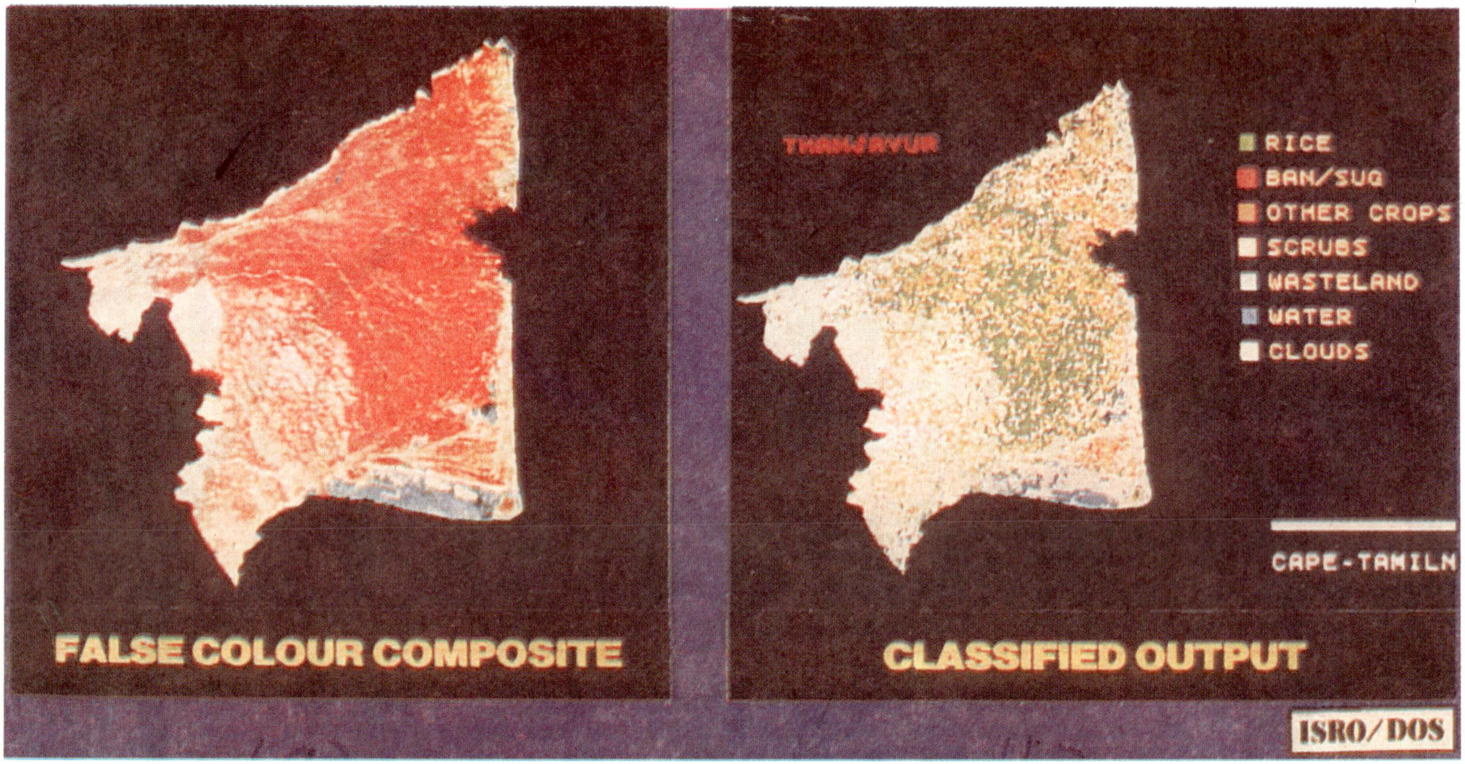

Plate 31.1 (a) and (b) Rice crop acreage estimation in Thanjavur, Tamil Nadu. (Courtesy ISRO and DOS)

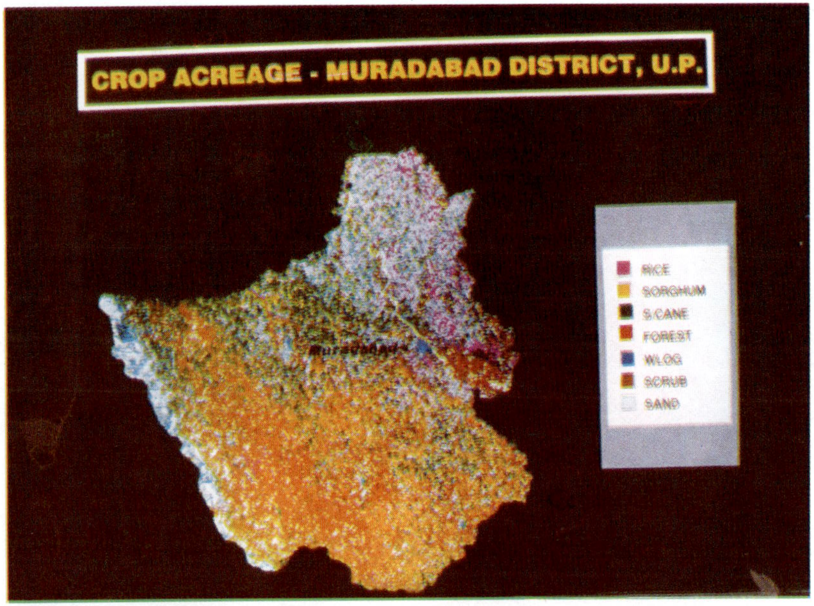

Plate 32.1 Crop discrimination, Muradabad District, UP, from IRS-1A satellite data analysis.

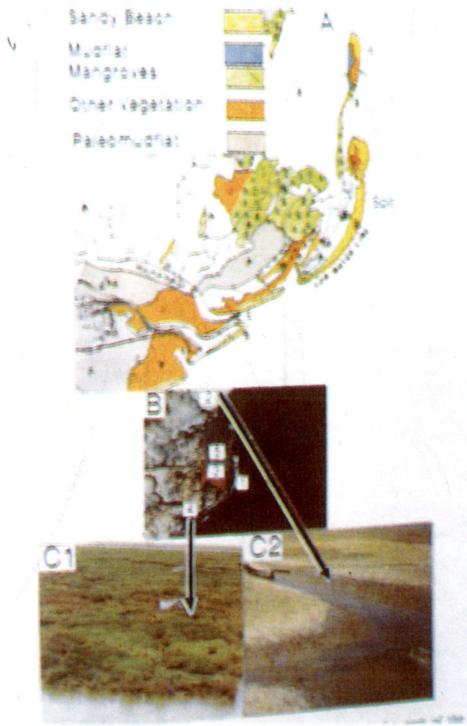

Plate 32.2 Coastal wetland map (A) shows a part of Orissa coast. Such satellite maps can be used to identify high tidal mud flats (C-2), mangroves (B-3), marshy vegetation (C-1) and bays.

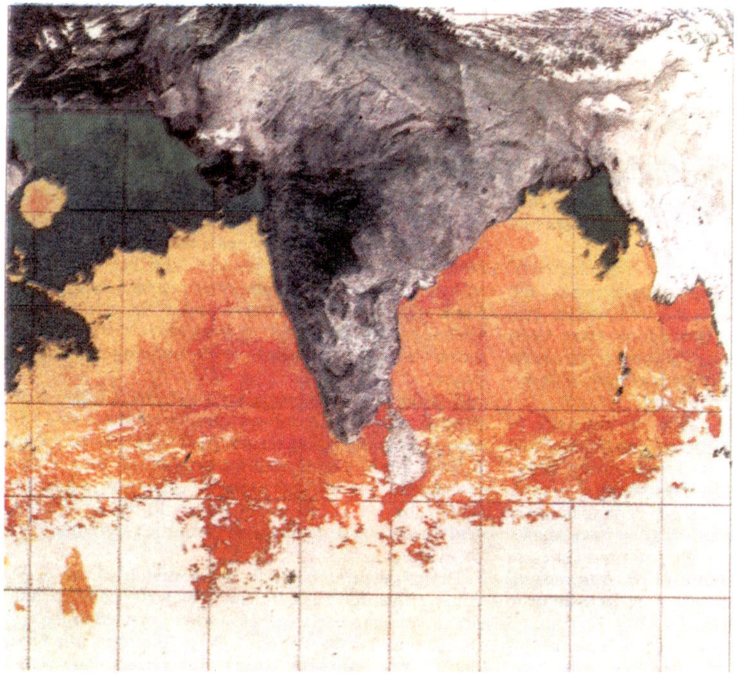

Plate 33.1 Colour coded image representing the sea surface temperatures (yellow 27°C, green 24°C and pink 29°C). Such maps can be of use to improve fish catch.

Plate 33.2 Digitally enhanced image of part of Ranga Reddy district, A.P. Such data will help classify features such as small fields and low lying areas.

Plate 34.1 Electronic distance measurement, Anna University.

Plate 34.2 Photogrammetry in action at Anna University.

Plate 34.3 Students undergoing computer training in IRS at Anna University.

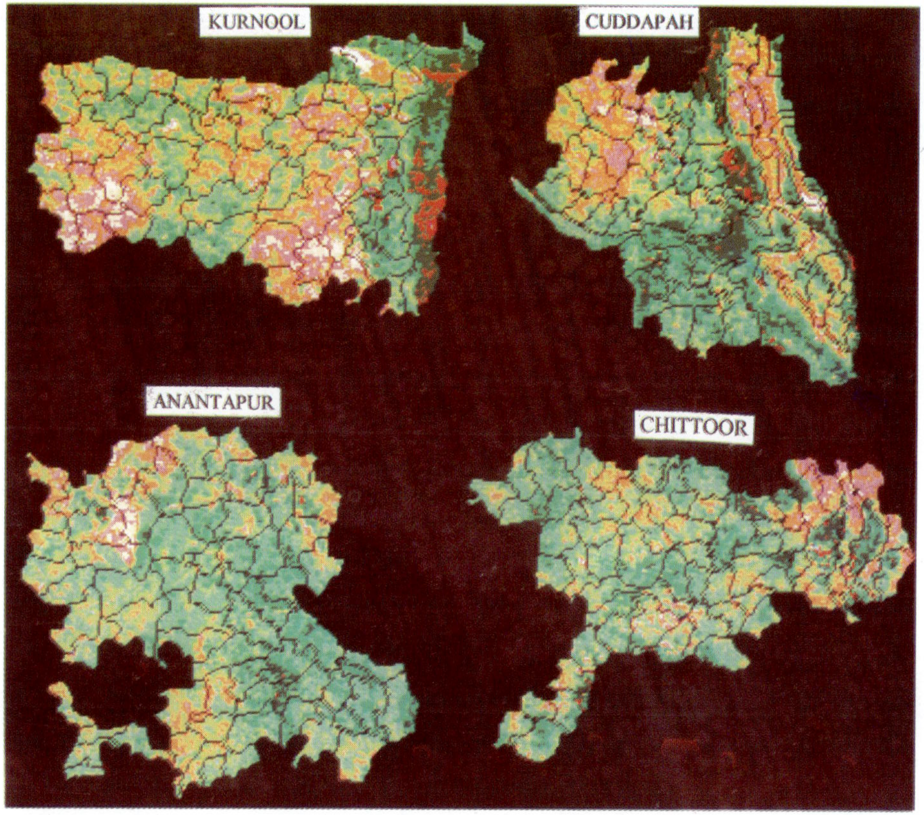

Plate 35.1 Detailed sub-district monitoring by NADAMS from a 1992 figure indicating mandal-wise vegetation index over Rayalaseema districts of Andhra Pradesh.

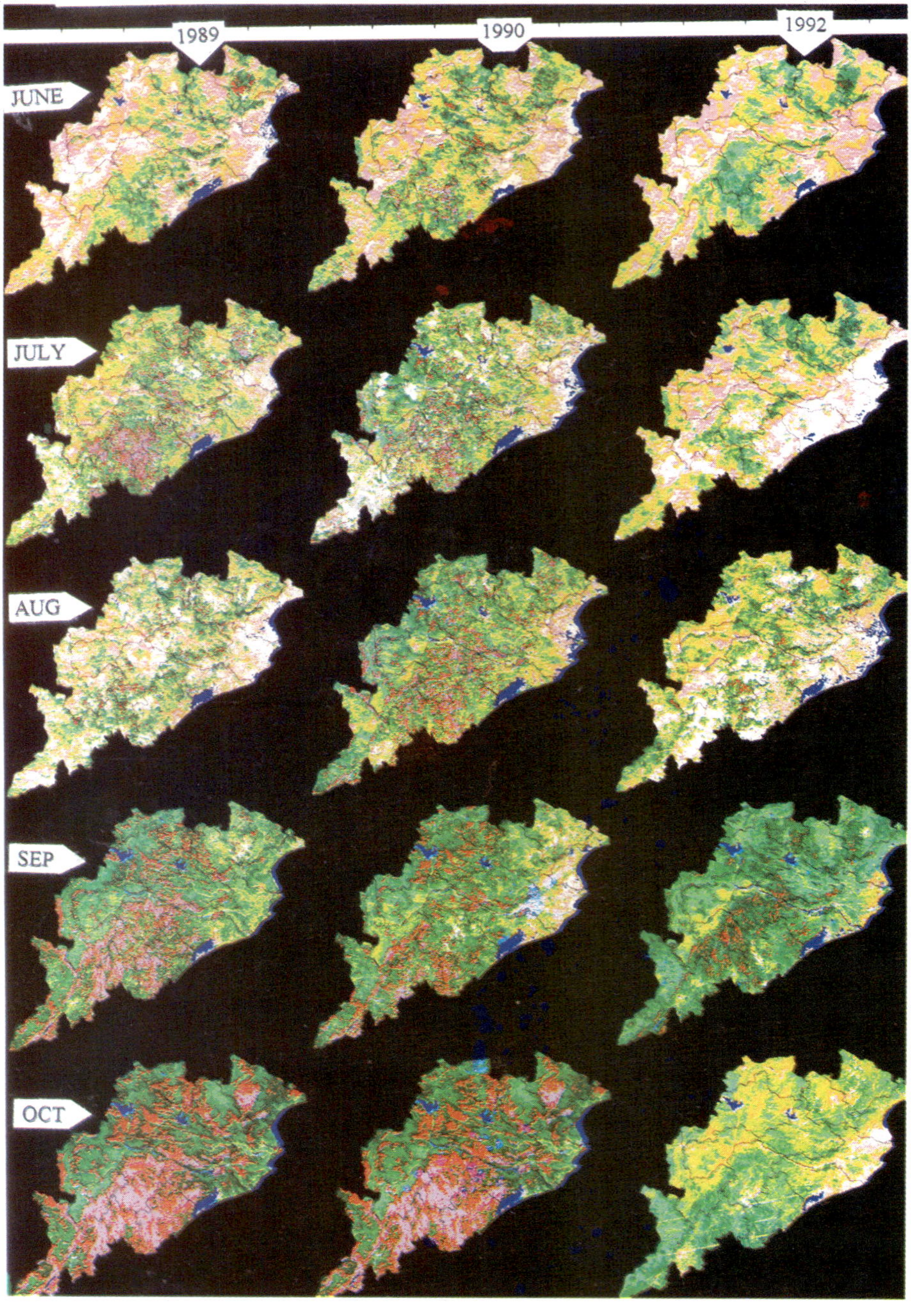

Plate 35.2 Monthly progress of drought situation in Orissa reflected in the 1992 vegetation index in comparison with 1989 and 1990.

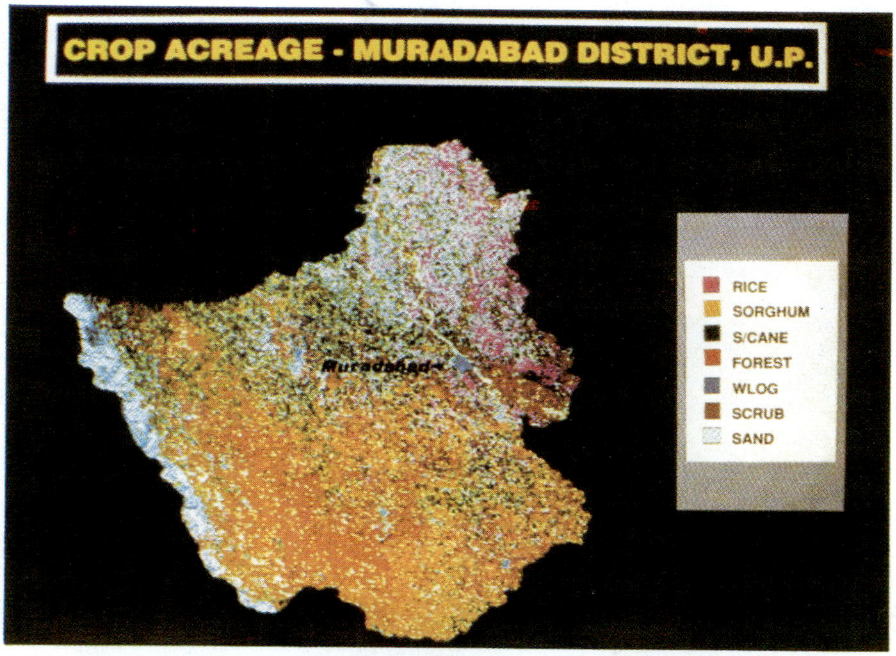

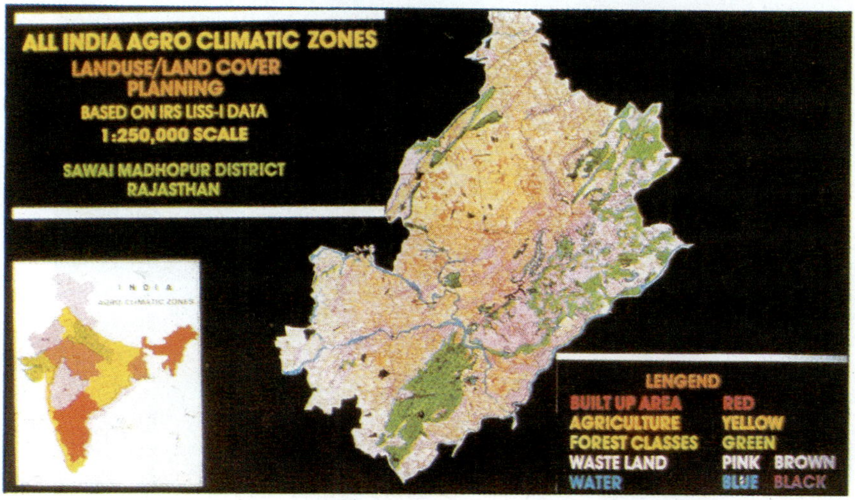

Plate 36.1(a) Discriminating cotton crop from other vegetation using remote sensing data.
Plate 36.1(b) Estimating pre-harvest acreage under cotton crop.

Plate 37.1 Classified map of Bhitarkanika National Park, Orissa. (Courtesy NRSA)

Plate 38.1 Image from RadarSat.

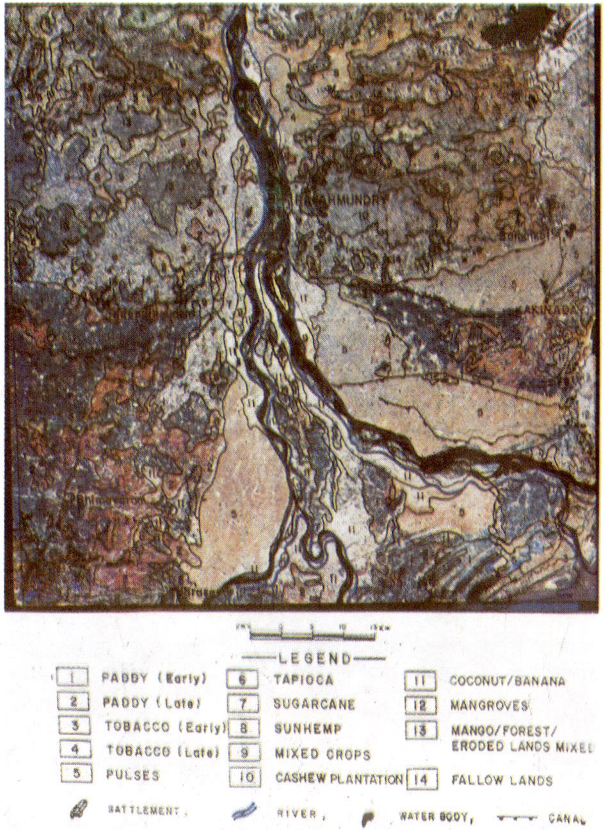

Plate 38.2 ERS-1 SAR data of Rajahmundry area.

Plate 39.1 Perspective view of Patnitop and its environs.

Plate 40.1 Perspective view with IRS.FCC dated 23-11-93 mapped over DTM.

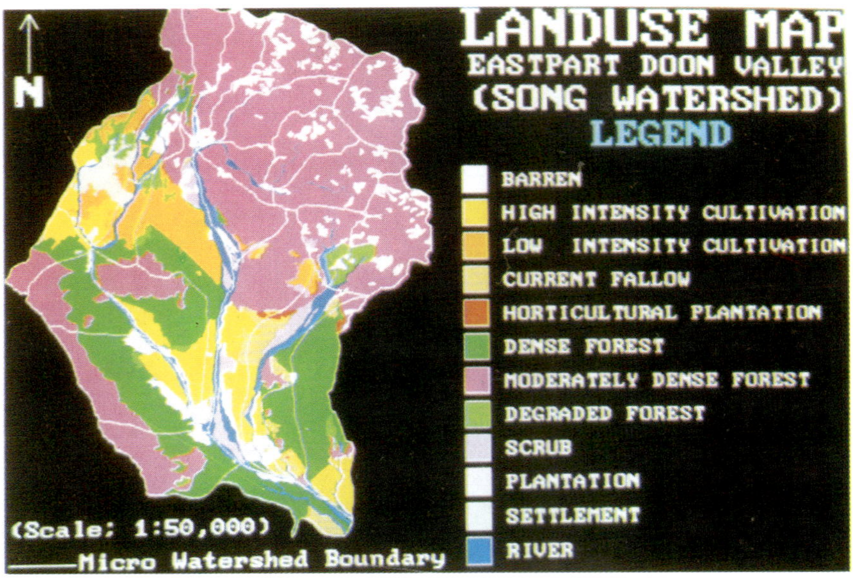

Plate 40.2 Land use map of the eastern part of the Doon Valley (Song watershed).

Plate 41.1 Identification of alignment of a pipeline for the Oman-India Sea route by the Indian Oil Corporation.

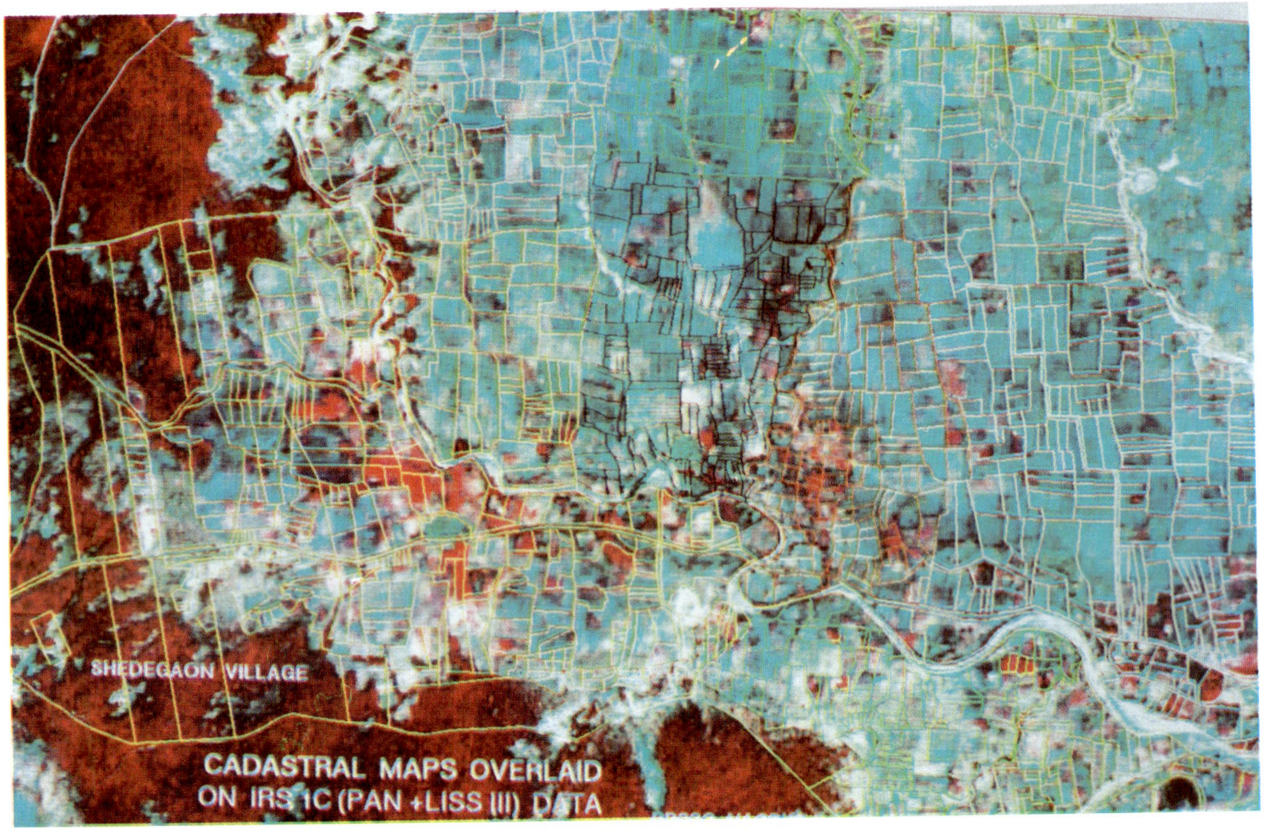

Plate 42.1 Satellite view of Shedegaon village and its environs. (Courtesy Maharashtra State Remote Sensing Application Centre, Nagpur)

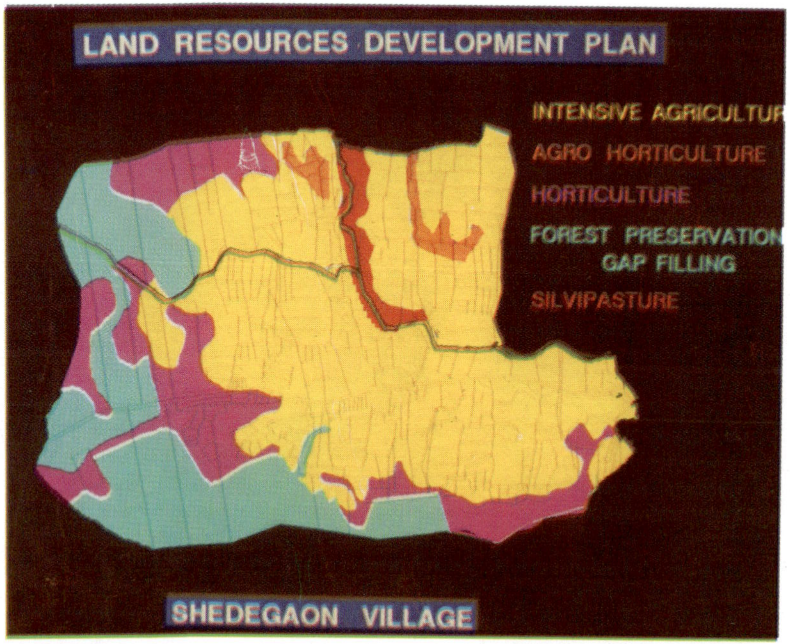

Plate 42.2 Shedegaon village: land resources development plan.
(Courtesy Maharashtra State Re .note Sensing Application Centre, Nagpur)

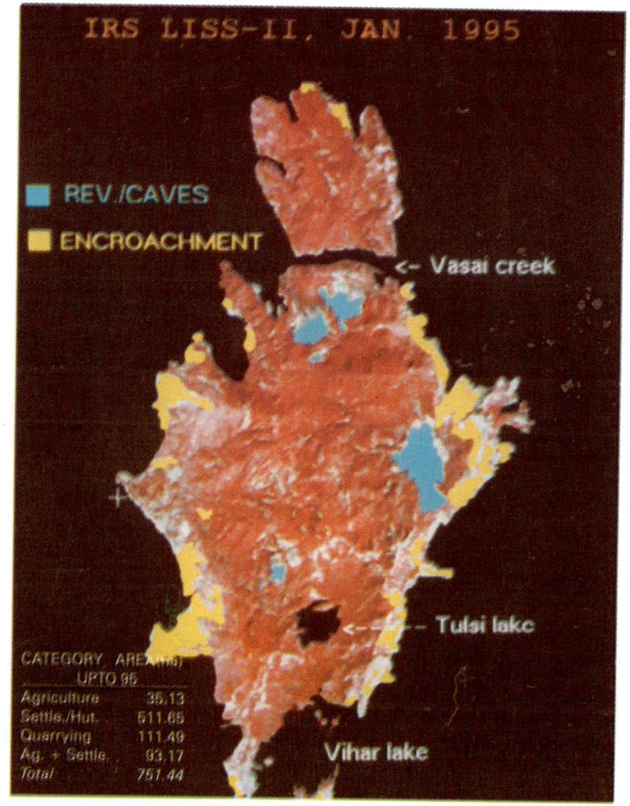

Plate 43.1 Encroachment studies for Sanjay Gandhi National Park.
(Courtesy Space Application Centre, Ahmedabad)

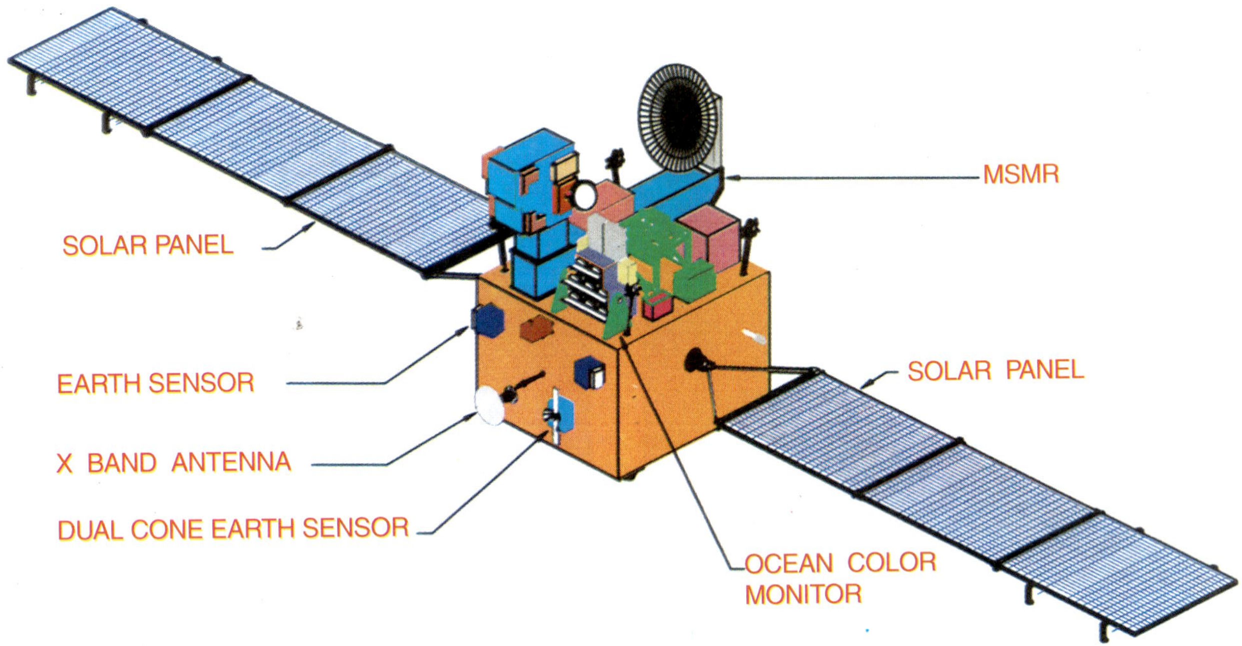

Plate 45.1 Deployed configuration of IRS-P4 spacecraft.(Courtesy ISRO Satellite Centre, Bangalore)

Study of the scenario of natural resources in and around the site proposed for the Bangalore international airport near Devanahalli, to establish the likely impact on the environment.

Some of the guide maps (of large cities) are on a scale suitable for tourists and other users and are kept updated periodically using IRS-1C panchromatic data in particular, with 5.8 m ground resolution as one of the main inputs.

An information system of Jodhpur city with a digital value-added package using IRS-1C satellite remote sensed data along with other ancillary information.

A study to select an industrial site at Nandikur in Dakshin Kannada, using remote sensed data as a main input, and several other similar studies using current remotely sensed data and the GIS, for a variety of planning requirements.

A recent and interesting example of the use of remote sensing and GIS technology was in the setting up of an industrial park in the Khatrej area of Mehsana district in Gujarat by the SAC (ISRO) in conjunction with Gujarat Infrastructure Limited, Ahmedabad.

A report released recently says: "The issue of industrial site selection is a classic theme of location theory. The site selection process resembles a real estate transaction in which industrial site agents deal directly with a client to find a suitable site that meets a series of locational and site-specific criterion.

The process is interactive—the client has a set of desirable site characteristics that include size, cost, utilities, existing built-up land and land use/cover pattern that can be coupled with a series of desired locational constraints. This can be developed as a multiple knowledge-based spatial decision support system by creating link between GIS and image processing techniques".

The Bangalore-Mysore infrastructure corridor reconnaissance and feasibility studies of possible routes for expressways and townships were carried out in a 2 400 sq km area along a stretch of 120 km between Bangalore and Mysore, and a 1 000 sq km area for a peripheral road linking two national highways on the east and west sides of the city.

This resulted in a successful selection process for development schemes, which would have taken a very long time had remote sensed data of appropriate quality not been used.

The main advantage derived in such studies is the fact that remote sensed data offers a near real-time and authenticated information of the study area and is amenable to digital processing so that planners can obtain much faster results.

A number of oil pipeline alignment studies for various oil companies have also been carried out to advantage using remotely sensed data. This includes identification of alignment of a pipeline for the Oman-India sea route by the Indian Oil Corporation (See Plate 41.1).

Needless to state here, what has not been easily possible for planning infrastructure development in the past, has now become much easier with the availability of the latest type of remotely sensed data in combination with other supplementary information and an integrated approach using the GIS and the GPS for more closely localised work.

Sometimes it is even difficult to predict and foresee how such data systems will become an essential component of all human activities and provide for better and sustainable living conditions.

The major emphasis in the Ninth Plan is on infrastructure development. So we have to find suitable technologies and practices to ensure the attainment of our cherished goal of moving closer to the global economy.

42

An Application of Geotechnology

Geoinformatics, or geomatics, can help boost agricultural yield. The article discusses this application and how the data can be disseminated at the field level by forming "information villages" and practising "smart farming".

Geotechnology is also called as geoinformatics or just geomatics, all conveying similar attributes and techniques used for gathering accessible information about the Earth—from the ground, aerial platforms or satellite platforms. While the first two are well-known, data from satellite altitude or what is generally termed as remote sensing is, comparatively, of recent origin.

Satellite remote sensing has made it easy for us to survey not only a range of our natural resources, covering large areas more economically, but also monitor the changes that take place. Initially, it used to be understood qualitatively, but as technology developed during the past two decades, a somewhat quantitative analysis also became a possibility because of other types of instrumentation being available, such as the GPS and advanced GIS. It is now possible, not only to integrate graphically displayed information precisely, tied to its geographical location on the surface of the Earth, but also statistical or socio-economic information, to simulate alternative planning data and use it for more realistic development activities. As the population increases, the per capita availability of land decreases, which means that ways and means of increasing productivity have to be found.

Essentially, geotechnology encompasses technologies like remote sensing, GIS, photogrammetry using aerial photography, GPS—for precise position location in terms of geographical co-ordinates—and other supplementary tools like electronic distance measuring instruments, digital theodolites, digital self adjusting levels, digital total survey stations, supplemented by computer aided cartographic and printing activities. In fact "traditional" survey techniques are still in vogue in certain specific cases, but the instruments have been improved to make surveys much quicker, easier and reliable. In fact, unless one is fairly knowledgeable and experienced, it may become difficult to make a judicious selection of the most economic methods.

While confining ourselves to how land management and agricultural productivity can be increased, it may be necessary to find all methods that are possible to do so, that may include the use of geotechnology. This is to get to know more about the physical character of the land besides factors like the soil, availability of water, the promotion of ideal seed usage and effective fertilizer and nutrient application. However, an area that may prove problematic is agroclimatic conditions and weather forecasting, particularly in areas which depend on rain for agricultural activities. India, being partly semi-arid, is dependent on ideal weather conditions that promote food production.

In an article, published by Dr M S Swaminathan, it has been pointed out that "micro-level management promotes the use of precision farming techniques involving plant scale agronomy rather than just area-based approaches. Plant-scale agronomy is knowledge and information intensive and affords opportunities for making farming intellectually stimulating in addition to being economically rewarding". The recent announcement of the three-pronged farm policy, emphasises an optimum use of land and water besides other inputs, although it is not clear how this can be done.

It is in this context that the use of geomatics or geotechnology in analysing information on land use is being attempted in some advanced countries. "Interest in precision farming has grown in the last five to six years—mainly in agricultural industries—in the US. Agricultural industries anticipate that a signified portion of their new business in the next two decades will result from providing new services and technology in precision agriculture," according to the Michigan University. There are lessons to be learnt from this. In India, large farmers utilise simple geomatics to carry out mapping as well as slope determination to improve production in areas where soil is weak in nutrient content. In this, targetted sampling may be guided by preliminary information about the site from remote sensing data, soil maps or terrain maps. A 50 by 50 m grid on a topographical map on a scale of say 1:2 000 or 1:5 000 or even 1:1 000 with contours at intervals of 25 cm or 50 cm may be the starting point.

To enable integration in a GIS software environment, all information may be needed in a digital format, which is not difficult as almost all field survey instruments show digital data. Cheap labour is another advantage in India. While some may argue that most small farmers may not understand such data, it is felt they can be trained to read numbers and figures on maps and use it. Once farmers get to know the layout of their farms, they will be able to evaluate the quantum of nutrient applications.

Unfortunately, we do not even have large-scale topographical maps available at the village and farm levels. Existing village maps or cadastral maps do not have height or slope information (See Plate 42.1). While the use of remote sensing or aerial photography may be thought of as a means to achieve this, it is felt that an appreciable amount of field operations may have to be supplemented so that data gathered can be integrated effectively in a computer with GIS possibilities.

Location-specific information is now becoming possible by the use of GPS and also laser ranging, which are ideal means of obtaining direct digital data. One may need a high level of expertise at the planning level, but at the operational level this can be achieved much more economically and made available centrally. In this context, the concept of an "Information village" a concept initiated by the M S Swaminathan Research Foundation, is worth emulating. In fact this will enable practical usage of geotechnology to improve agricultural practices. However, these measures cannot he viewed exclusively in the area of agricultural research and development activity, but should form a part of it, to obtain optimum results.

In fact, what is crucial in precision agriculture is having a map of spatial patterns and physical dimensions that identify locations and the aerial extent of distinct management zones within the field. It is called "smart farming". The idea involves the judicious usage of remotely sensed data. The better the spatial resolution, the better the results of targeted information (See Plate 42.2). Perhaps the next remote sensing satellite with a 2.5 m ground resolution, likely to be launched by the Department of Space, may be used extensively in the area of application.

"Agricultural practices are changing in response to economic technological, environmental and social trends. The profitability and the environmental impact of a farming system or systems are of concern to both farmer and the government. Precision agriculture, which takes into account spatial variations of soil at field level, is being proposed as a remedy to many of these environmental problems". The focus of this article is on the effective and judicious usage of geotechnologies to reach targets such as relating agricultural operations with the information derived from the economic usage of such technologies.

It is a fact that there are newer methods now for economic farming. This is in addition to obtaining better agricultural yields using precision techniques besides enabling improvements in a variety of other parameters that affect agricultural practices. It may be prudent to discuss some of these aspects of geotechnology at the field level at symposia to empower farmers with more information and knowledge.

43

Estimating Forest Cover

This paper looks at the applications of remote sensed data in forest management.

Remote sensing by satellite became a reality in the early Seventies with the launching of the first earth resources technology satellite (ERTS) by the US. It was later renamed as Landsat 1 as satellites of this series went into orbit over a period of time. Several other countries also launched remote sensing satellites with types of sensors and different spatial and spectral resolutions.

An article on the use of remote sensing for assessing forest cover, a result of a study carried out by the NRSA, Hyderabad, in 1983, illustrated the reduction in forest cover from 1972 to 1983. With the available satellite remote sensed data obtained then from the Landsat satellites having a ground resolution of about 80 m, covering the entire country, it was found that the country had lost about 2.8 percent of forest cover in about seven or eight years time—the area classified as closed forest (which means forest area having crown density of more than 40 percent), had suffered the most. The closed forest area then stood at about 11 percent of the total geographical area of our country.

Being the first attempt, the Government started the Department of Forests and Environment to safeguard forest cover. The Forest Survey of India (FSI) finally adopted the use of satellite remote sensing methods as a means to monitor our forest wealth every two years. As time went by, better quality satellite remote sensed data also started arriving. Now we have the world's best satellite remote sensed data from the remote sensing satellite series—IRS-1A, IRS-1B, IRS-1C and IRS-1D besides the P-2 and P-3 satellites. In fact, the world community has acknowledged the quality of our data and is using it in a range of applications.

In spite of all our efforts by various State forest departments, it has been documented that since 1995 the country has lost 5 500 sq km of forest cover according to the Union Environment and Forest Minister, Mr Saifuddin Soz, on February 7. It was also announced that the ministry plans to conduct surveys in Andhra Pradesh and Madhya Pradesh—which have lost 3 822 sq km and 3 969 sq km of forest area. Based on satellite data for the period 1993 and 1995, the forest cover of the country was estimated at being 19.27 percent of the total geographical area, an appallingly low figure. The area under dense forest has been estimated to be 11 percent. However, in Maharashtra, there has been an increase of forest cover over an area of about 2 300 sq km.

While there are various reasons being attributed to the change in forest cover, we do have an effective mechanism to monitor the changes as is now being done by the FSI. Some experts feel that the method of classification of forest cover based on the criterion of density should change, and a more detailed system by splitting 40 percent of crown density of defining what a closed forest is, should be attempted. However, it can be stated that satellite remote sensing data is improving in terms of both quality and quantity, particularly in terms of spatial and even temporal resolutions.

The Forest Survey of India at Dehra Dun, is well equipped to make the best use of remote sensed data as well as evolve useful information based on up-to-date software systems established five years ago. Among its staff members are forest officers, scientists and computer engineers.

The State of Forest Report, 1997, published by the Forest Survey of India, is an excellent document, giving a detailed analysis, with illustrations, of changes in the States.

One reason for a reduction in forest cover is illegal felling; another reason could be encroachment; yet another is attributed to tribal communities practising slash and burn cultivation. This is prevalent in the North-East of India. People do not even spare reserve forests. A typical example is the Sanjay Gandhi National Park, Borivili, on the outskirts of Mumbai. Studies by the scientists at the SAC of the ISRO, has shown that the park, spread over 103.09 sq km, has a wide range of habitat conditions, ranging from mangroves to evergreen forests. This study was carried because of serious problems of encroachment. The scientists used data from the French SPOT satellite (1990), the Indian remote sensing satellite LISS II (False colour composite) (1994 and 1995), the Sanjay Gandhi National Park management map on 1 inch to 1 mile scale, and the Survey of India topographical map on a 1:50 000 scale (Plate 43.1).

With the data, the Maharashtra Forest Department was able to win a legal battle against encroachers. In fact, the value of the area under encroachment would be worth several crores of rupees today. This is an example of the use of such remotely sensed data in providing legal evidence.

Forest Fires: The loss of forest cover because of fires, especially in the tropical countries has been alarming. Sometime ago, I had written an article showing how a typical forest fire in Karnataka was studied using satellite remote sensed data, and the loss in quantitative terms analysed. While forest fires may continue to occur, it may be possible for forest officials to make quick decisions in fighting it using such data.

The recent forest fires in Indonesia crippled life in Malaysia and Singapore. The loss of forest wealth was analysed using remotely sensed data and photographs taken by micro-light aircraft fitted with cameras.

The amount of money spent on these measures would be a fraction of what we regularly allocate and expend for various forestry schemes, annually. It is heartening to note that the Tamil Nadu Forest Department is using remote sensing data and GIS technologies for a variety of forest management activities. It may also be advantageous to develop an integrated forest fire management system in the coming years in co-ordination with all state forest departments, on the lines of the one evolved recently by a company called the Earth Observation System. The concept of an Integrated Forest Fire Management System (IFFMS) is worth exploring too.

Thus, it can be concluded that many technologies have already been developed or are still being developed, particularly in the area of satellite observation combined with digital technologies. It will be wise to make use of these to our benefit, particularly where the facilities have already been established.

44

Tapping Potential

A good GIS revolves around the effective use of proper database management, computer-aided design and the use of up-to-date information systems such as remote sensing. The potential is great and it may be the only way that our country progresses in terms of agriculture, effective use of water resources, forest management and infrastructure.

On October 7, last year, there was a seminar, "Map India 1998" organised by the Centre for Spatial Database Management and Solution, which publishes a magazine called *GIS Development*, located at NOIDA, near Delhi. People involved with GIS attended this seminar. Several large information technology companies exhibited their products. The seminar was co-sponsored by reputed organisations, the prime sponsors being Tata Infotech, the Indian Space Research Organisation (ISRO), Autodesk Inc., the Institution of Engineers (India), one of the largest engineering professional forum, the National Atlas and the Thematic Organisation (NATMO) and the National Remote Sensing Agency (NRSA). This seminar was an eye-opener as to why GIS is not picking up speed in our country, although it is considered a versatile tool all over the world for several applications and planning for development. The participants highlighted the deficiencies in development management and progress.

There is no lack of knowledge in GIS and its applications. The main deficiency in using it in micro-level planning and as a decision support system for towns and rural areas is the lack of a reasonably accurate geographic, location-oriented, large-scale cartographic database on which other significant variables could be built. The largest cartographically accurate database now available is in the ratio of 1:25 000 or 1:50 000, which can be extracted from the topographical maps of the Survey of India. This is grossly inadequate for micro-level applications.

For micro-level watershed programmes and improving cadastral information, a reasonably good geographically accurate cartographic database on the scale of 1:10 000, 1:5 000 and even 1:2 000 is needed. Besides, we do not have slope or relief information in the form of contours at say 1 m or less, or spot heights provided at regular intervals for an effective implementation of GIS, particularly for improving agriculture productivity, economic use of inputs and conservation of water. Many organisations are still to address this problem; so far they have been saying that the use of 1:50 000 topographical maps or 1:25 000 scale maps will serve this purpose. This, I feel, is incorrect and does not serve the purpose especially with regard to relief information. For example, a farmer needs his field information in terms of its physical characteristics, besides other parameters such as soil conditions for making maximum use of his land to be more economic and sustainable. Because such areas are rather small, the use of aerial photography, even on a large-scale, may not serve the purpose and be economically viable. This is because it is rather expensive and is under the control of the Ministry of Defence, as classified information, to be used only by the government.

Secondly, producing relief information using aerial photographs at a level interval of say a metre and below is not easy, and one has to go to the ground to obtain it by using modern auto-setting levels. Similarly, for mapping towns and using such base maps for facility management and automated mapping (AM/FM), we have to have accurate base map information in digital cartographic environment, as otherwise a good GIS or LIS (Land Information System) is impractical for day-to-day use and management. There will be several users for such a system, such as planning and executing water distribution management, sewerage planning and management, electrical power distribution network, telephone cable distribution network, transportation network, and several other facilities. While this concept is recognised, perhaps it is not known how a good digital cartographic database with at least a network of spot heights could be produced.

Except for organisations like the Survey of India, others may not have a standard system and practices that are required for this. In view of the fact that the Survey of India, being a very large organisation and not meant to provide such an information base, it has to be carried out by others with the support of organisations in the private or semi-private sector with adequate manpower and machine power facilities. Relief or height information at a detailed level can now be done only by ground-levelling based on mean sea-level benchmarks, which are generally available in most places, as provided by the Survey of India. A reasonably accurate base map in rural areas could be drawn up at a reasonable cost and in a short time if the area is small, say 1 000 to 20 000 ha per team.

A number of such teams can be entrusted with such tasks, generally managed by experts, who understand the pitfalls that may occur in carrying out such surveys and can take corrective measures if necessary.

This method may appear expensive and time consuming, but it is not so. In our country, manpower is cheap and the use of modern field instruments with digital outputs can expedite this activity and be computer compatible. In addition, our outdated cadastral maps can be corrected and made to fit in with such large-scale maps and even stored in digital form, which will make the task of incorporating changes or periodic updating extremely simple. Training existing staff in this field will not be very difficult.

A good GIS revolves around the effective use of proper database management, computer cartography, computer-aided design and the use of up-to-date information systems such as remote sensing. The potential is great and it may be the only way that our country can progress in terms of agriculture, urbanisation, effective use of water resources, forest management and infrastructure. Let us not underestimate its worth and feel that it can be done only if computers and other peripherals are made available. There are several digital as well as non-digital equipment and instruments available, and making a suitable assessment with a clear objective of the possibility of making available large-scale digital cartographic database is not impossible. Nor is it too time consuming or uneconomical, in terms of improving our LIS and cadastral update.

In this context, the recent Map India 1998 conference was an eye-opener to both users and vendors in the GIS/GPS fields. The lack of data and a reluctance in sharing data or information available with different people or organisations, undue secrecy and restrictions prevailing in the usage of certain types of information and data, a lack of understanding of its potential at the political and bureaucratic levels and a will to adopt it as it is, or adopt and use it are some of the major reasons against making GIS/GPS acceptable. Its use in industry, census departments, elections, revenue intelligence, tax collections, natural resources assessments, monitoring, updating and effective management has to be highlighted aggressively.

This article is being written to enable the public to know why certain technologies are not easily picking up and what action can be suggested to improve the situation. About a few months ago there was an Information Technology (IT) task force constituted to formulate the draft of a National Informatics Policy, chaired by Mr Jaswant Singh, Deputy Chairperson of the Planning Commission, co-chaired by Mr N Chandrababu Naidu, Chief Minister of Andhra Pradesh and a number of members. There were 108 recommendations, touching all segment of the IT sector, including GIS In view of the work involved, it also recommended making full use of the private sector.

While the Survey of India, during the last five or six years has developed digital cartographic database on a scale of 1:2 50 000 it is still not open to the public to produce digital cartographic data base on a larger scale of 1:10 000. This is a massive operational endeavour as far as India is concerned. Unless our policy on restrictions and security aspects are reconsidered and a large number of private entrepreneurs come forward to carry out the task, GIS for our various development programmes, particularly for spatial planning, will only remain a topic of discussion.

In this context, a review of the Official Secrets Act, the proposed Freedom of Information Act and the implementation of the recommendations made by the IT Task Force is necessary.

During the Map India Conference the recommendations made were:

i) a national task force on IT to address the problems faced by the GIS community more significantly.

ii) the implementation of Right to Information Act as an important way in which the GIS group can get over the problem of scarcity of data.

iii) every government organisation must include "data dissemination" as a part of its mandate. The internet should be used to achieve this objective.

iv) the government and industry must work towards making software and the databases more accessible and affordable.

v) a need to create a meta-database of spatial and non-spatial databases available in India.

vi) an urgent need to develop standards for the GIS database.

As pointed out, increased manpower will have to be deployed, particularly into the field, equipped with modern digital instruments to acquire data which may serve a number of users.

In conclusion, let us hope that the government realises the importance of getting into the fast track of progress, by doing whatever is required, without any more procrastination or reservations.

45

Understanding Oceans

The recent launch of the IRS P-4 satellite provides us an opportunity to study ocean resources and its characteristics and see how best we can benefit over a period of time, in several areas of human survival, specifically related to food security on a sustained basis. With this, we have come a long way in the last two decades in making our own remote sensing satellites.

The successful launch of the IRS-P4 satellite at Sriharikota on May 26, 1999, a fully indigenous effort, will help us know more about our oceans and their wealth (See Plate 45.1). As a result of the law of the seas coming into effect, a distance of 200 nautical miles from the coast-line off-shore into the ocean forms the exclusive economic zone. This is of particular significance to India, for it has a long coastline, about 7 500 km, and covers an equivalent of two-thirds of land area. The P-4 satellite provides us a challenging opportunity to study ocean resources and its characteristics and see how best we can benefit, over a period of time, in several areas of human survival specifically related to food security on a sustained basis.

We have come a long way in the last two decades in fabricating our own remote sensing satellites and the Indian Remote Sensing satellites 1A and 1B were the first two operational satellites which were launched nearly a decade ago. Then came the next generation satellites, IRS-1C and 1D, which helped us study our natural resources for better management and provided greater insight into the intricacies of fabricating such satellites. As of now, these IRS-1C and -1D are known as the best land resources sensing satellites for civilian uses world over.

In the meantime, our developmental efforts to fabricate polar satellite launch vehicles (PSLV) progressed rapidly. Although the first PSLV carrying a small satellite, named IRS-P1, could not succeed, the IRS-P2 satellite launched by an indigenously fabricated PSLV went into orbit beaming data similar to the data available from IRS-1A and -1B on October 15, 1994. This satellite was put into orbit at a height of 817 km, with a 24 day repetivity.

This firmly established our confidence in being self-reliant, and the successful launch of a IRS-P3 and ID, and now the IRS-P4, which has only confirmed our scientific and engineering abilities and skills which very few countries possess.

A brief note on the IRS-P3 satellite which went into orbit on March 26, 1996. This satellite launched by PSLV-D3, at an altitude of 817 km, had an astronomy payload and two remote sensing payloads, a Wide Field Sensor (WiF) and Modular Opto-Electronic Scanner (MOS), essentially catering to certain oceanographic applications.

The modular opto-electronic scanner (MOS) is primarily meant to provide data for ocean-related studies, coastal water discrimination, sediment transport, ocean dynamics and littoral drifts. The MOS sensor has three modules, MOS-A, -B and -C, each having varying resolutions and varying spectral bands, to enable specific applications. The Wide Field Sensor (WiFs), capable of viewing a swath around 770 km, with a spatial resolution of around 188 m, has specific applications and understanding of vegetation dynamics on a continuous basis. This sensor is nearly similar to the coverage from IRS-1D and has been used very effectively where a large area has to be studied and understood in respect of vegetation status.

The one launched recently has special significance as it is the first of its kind, primarily devoted to our ocean parameters, apart from the fact that the same PSLV launched two other satellites, KITSAT-3 of Korea and DLR-TUBSAT of Germany, all orbiting at an altitude of 727 km in a Sun-synchronous orbit.

The IRS-P4, called the Oceansat, has on board two main sensors, the Ocean Colour Monitor, (OCM) and Multifrequency Scanning Microwave Radiometer (MSMR). The ocean colour monitor is an eight channel sensor, operating in the visible and near infrared region of the electromagnetic spectrum. It will have a swath of 1 420 km, and a ground resolution of 360 m along the track and 246 m across the track. It has a repetivity of two days; details of the various spectral bands are not being elaborated here at present.

The main applications of the ocean colour monitor are to monitor and indicate potential fishing zones in coastal waters, exploration of deep-sea fishery resources and other related information. It can also provide data on coastal processes, such as sediment dynamics, dynamics of estuarine/tidal inlets, prediction of shoreline changes, circulation and dispersed patterns, littoral drifts occurring in certain coastal regions, marine pollution and oil slicks, coral reef studies, etc.

The MSMS sensor, will be providing information on the monsoon corrective systems and help us in understanding the meteorological conditions. The utilisation programme of the data made available by IRS-P4 has been well co-ordinated by involving agencies such as the SAC, ISRO the Meteorological Department National Institute of Oceanography, Goa, the National Centre for Medium-Range Weather Forecasting, the NRSA the National Physical and Oceanography laboratory and various universities.

I would like to say something about the future satellites. The next in the list is called IRS-P5 (Cartosat-1) data from which is primarily aimed at cartographic applications, i.e. to make accurate maps of possible topographic scales and to update existing topographical maps more frequently than what is done at present.

It is expected to be launched at an altitude of 617 km, in about two years. It will provide data at a spatial resolution of 2.5 m with a 30 km swath. Provision has been made to have stereoscopic viewing possibility as in the case of aerial photographs. The weight of this satellite is expected to be 1 500 kg. The data provided by this satellite would be of special relevance to our country to develop high quality GISs for more detailed micro-level planning programmes.

The next in line is the IRS-P6, called Resourcesat, which is mainly planned to provide data focussing on agriculture and water resources, land use studies and planning, soil mapping, forestry and geologically related applications. The data may be similar to IRC 1C/1D, except that it will have a 5.8 m resolution multi-spectral band and with stereoviewing possibility. This may be launched by 2002 and may have a follow up launched once in five years or so with minor modifications.

All this indicates that we are now on a path of not only attaining self-sufficiency and providing continuity in remotely sensed data, but also that we are on the way to understanding our natural resources better, to enable efficient and sustainable management.

46

Managing Groundwater

Replenishing groundwater is essential to ensure sustainability. Remote sensing serves to locate areas where artificial recharging of ground water can be carried out. What should be kept in mind is that water should be extracted only to the extent that it can be recharged.

Our water resources are fast getting depleted and the time is not far off when water will have to be rationed. A large part of the country falls under the category of drought prone regions. Rainfall is limited and occurs in a very small period of time in a year, so as much rainwater as possible has to be conserved and our groundwater potential has to be enhanced. Water is a renewable natural resource and its usage needs to be well managed to ensure sustainability. This is particularly important with respect to groundwater which, if over-exploited, may lead to disaster. The water extracted should be replenished by all available methods. At present, the groundwater level in many areas is going down as replenishment is not keeping pace with extraction. Several methods have been practised from time immemorial to ensure that as much rainwater as possible is stored in tanks during the rainy season, making it possible for as much water as possible to percolate below the ground, so that it could be extracted when required. However, this is not possible in all areas as the geological formation varies from place to place and a lot of water stored on the surface in tanks and the like evaporates during summer.

A large population in rural areas depends on groundwater. Nearly 60 percent of the total irrigation potential comes from the use of groundwater. The distribution of underground water is not uniform and its availability is subject to wide spatio-temporal variations depending on the underlying rock formations, precipitation patterns and other hydrological conditions. It is therefore essential to control and manage the groundwater available in a scientific manner. If we exploit this water without keeping track of the amount of recharge that takes place year after year, very soon we will find that the level has become dangerously low. Regulated extraction is therefore essential.

Groundwater movement and occurrence are generally controlled by the geological structure and land-forms and their hydrogeological characteristics. By integrating information about rock types, geological structure and land forms, we can prepare hydrogeomorphological maps, which will give us an idea of the potential zones where such resources could be available. For this type of work, remote sensed data from satellites have been found to be very useful and a quick means to obtain baseline information about the occurrence and movement of groundwater, including recharging conditions.

The usefulness of satellite data is in locating fractures or faults that show up as linear features, which usually are the zones of localisation of groundwater in hard rock areas and other favourable features. Remote sensing technology has been extensively used in a GIS environment for about two decades and has resulted in successful exploitation of groundwater in many areas. Unfortunately, it has led to over exploitation of water, particularly for irrigation, and the level has been going down every year. In many places the water has become brackish as there is not enough recharging by natural processes or by artificial methods. In urban areas, especially near the coast, ground water has become brackish, and has led to a situation where people have to buy water. Hence, we need to find areas where artificial recharging of ground water becomes necessary and at the same time adopt measures to extract water only to the extent it can be recharged when precipitation occurs. This may be difficult to begin with, but it is not impossible. For identifying such areas where artificial recharging is possible or attempted, the use of remote sensing and GIS becomes an essential tool.

Realising this, the Tamil Nadu Government is identifying areas with scope for artificial recharge, making use of remote sensing and GIS and the facilities available at the Institute of Remote Sensing, Anna University, on a project mode during the past two years or so.

The success of this effort will show up only after a few years and it will go a long way in enhancing the water potential of water-starved Tamil Nadu, and can be a lesson other states with a similar predicament.

Earlier, remote sensing helped identify potential areas to target and extract groundwater. Now, the same technology serves to identify potential zones where artificial recharging of groundwater could be undertaken more effectively and scientifically. To some extent this could also enable monitoring the extraction of groundwater keeping in view the amount of recharge, so that the concept of sustainability is maintained. A construction company in Chennai has been promoting water conservation. It has distributed pamphlets detailing how rain water could be prevented from being wasted and instead made to seep into the ground.

A year ago the Institute of Remote Sensing, Anna University, was asked to locate zones where artificial recharging measures could be undertaken. So far 386 blocks in Tamil Nadu have been covered.

The various aspects considered were: the geomorphology, depth of weathered zone, hydrological soil group, slope, run-off, geology, land use, lineament density, general water level below the ground at any given time, rainfall, drainage density and water quality. Availability of remote sensed data and topographical maps were a primary source of data for obtaining most of the information. Their recommendations were construction of recharging structures like percolation ponds, check dams, *ooranies* and recharge pits. Block-level maps were prepared on a scale of 1:50 000 and maps for each zone indicating favourable, moderate, less and poor zones, by integrating various thematic layers using computers with the help of GIS software.

Elaborate work was carried out by special teams on an urgent basis. If this effort, probably undertaken for the first time in such an elaborate way by the Institute of Remote Sensing, Anna University, proves to be efficient then it could be repeated in other states which are short of ground water. It may be concluded here that remote sensing, which helped in a big way to locate potential areas of ground water availability, has also been found useful to locate zones where artificial recharging of ground water could be carried out in a big way. This only shows the versatility of this technology. Such efforts should not be considered as a one-time attempt, but have to be continued periodically to monitor and balance the extraction versus the recharging, to ensure sustained availability of this resource for a long period of time.

47

Databases for Geographic Systems

With the increased possibilities of connecting geography and development, and with the evolution of geomatics-based information technology, spatial or geographic information is being used in all aspects of Earth's resources assessment, management and monitoring. Therefore, creating cartographically accurate databases for developing geographic information systems to meet the needs of micro-level planning is a must.

This article will deal with the importance of a geographically accurate cartographic database which would help create a reliable GIS. The result of this will be that planning processes as well as decisions arrived at will be accurate with respect to the geographical location of a subject.

Remote sensing gives us information about the natural resources available on the surface of the Earth at periodic intervals as well as monitors changes. While information on remote sensing ensures up-to-date data of various natural resources—particularly renewable resources like water, agriculture, forest and land use—it must be noted that such information has to be used in bringing out an integrated information system.

To ensure a good and reliable GIS, base details available from an existing map, (the scale is generally 1:50 000 or 1:25 000) are used. Not many of us may know or understand what goes into the making of an accurate map on a scale suitable for use to produce an accurate and reliable GIS. This is because the Survey of India maps are available.

Such a map (on a 1:50 000 scale) can be digitised and used as a base-map for any GIS based activity. But it defines the position on the surface of the Earth, only to the extent of what a small dot on such a map would represent on the ground. In this case, for a position extracted from a 1:50 000 scale topographical map, a small dot of less than a millimetre may represent a dimension of about 12 to 15 m or even more on the ground.

In some instances, the inaccuracy of such a magnitude on the ground may not be acceptable. Scientists feel that satellite remote sensed data of a higher spatial resolution will solve this problem, but it should be realised that it is not possible and will lead to appreciable inaccuracies. The cartographic database is like a foundation on which other data is built.

At the annual convention of the Indian Society of Remote Sensing recently, which dealt with "Remote Sensing Applications for Mapping from Space: Retrospective and Perspective", the Chairman, Indian Space Research Organisation (ISRO) and the Secretary, Department of Space, Government of India, said that inspite of advances in satellite remotely sensed data, most mapping is still based on aerial photographs. But why?

Aerial Photography

Aerial photography, though expensive, ensures geographically accurate maps of different scales. Even today, photogrammetry, defined as the art and science of making accurate measurements on aerial photos, is not only valid but is also rapidly advancing from the analytical mode to the digital mode or soft copy photogrammetry.

When remote sensed data from satellites was made available after the launching of the first Landsat satellite by the US in 1972, it was felt that aerial photographs, as a tool in map-making and in the interpretation of resources, would be phased out. However, after nearly three decades, we still make do with aerial photography. It is now a system that uses both modes.

When the requirement of an urban environment is taken into account, maps on a much larger scale than 1:50 000 or 1:25 000 are needed. Enlarging such maps beyond a certain limit will only enlarge the errors. Similarly, for agricultural operations and micro-watershed planning, maps of scales around 1:5 000 or 1:10 000 are preferred along with contours of at least 2 m intervals or even closer, which satellite data cannot produce. However, when high resolution data of say a metre or so becomes available in the course of time, it may serve a purpose, but to what extent it can be compared to maps brought out of systematically flown aerial photographs is yet to be proven.

The first 1 m spatial resolution satellite, 'IKNOS', launched by an American company recently, was lost in space. One must also remember that when higher resolution satellite data becomes available, the coverage per scene will be smaller and the cost higher.

One of the shortcomings of satellite remote sensed data, till about a decade or so ago, was the lack of stereo viewing, or seeing land data in three dimension. However, India has succeeded in obtaining such data (using satellites IRS-1C and -1D). It is only hoped that the next satellite, "Cartosat", to be launched after a year or so, will have a 2.5 m spatial resolution data with stereo capability.

Infrastructure: The infrastructure to carry out mapping from aerial photography is limited to the public sector. There are a few private sector organisations which have established such facilities and which are able to carry out mapping activities using the latest analytical and digital photogrammetric techniques at a reasonable price, and under supervised security. Speck Systems Ltd., Hyderabad, which did mapping for a foreign country, worth a few million dollars, has the ground equipment needed and cartographic skills to produce large scale maps of acceptable accuracy.

It is clear that satellite remote sensing data, by itself, cannot match the data from aerial photography and in providing large-scale accurate cartographic databases that can be used in micro-level planning and micro-level GIS usage.

The Future: With the need for increased food production, we have to develop and bring these technologies in association with GPS and ground-based surveys. Ground-based surveys use modernised instruments which provide digital data that can be fed into a computer.

With the increased possibilities of connecting geography and development and with the continuing evolution of geomatics-based information technology over the last two decades, spatial or geographic information is increasingly being used for strategic decision-making in all aspects of Earth's resources assessment, management, and monitoring. In fact, such information technologies are of importance to environmental and natural resources information management in all developing countries.

Policy makers should realise that creating large-scale cartographic databases is a one-time effort. We should also realise that while analysis of computer data may look easy, the process of translating analysed data to reality is not as easy as one might imagine. A uniform 1:10 000 scale cartographically accurate database must be created with a contour interval of say a metre or two immediately, for developing a GIS to meet the needs of micro-level planning. Perhaps there are private enterprises which may be willing to invest in aerial photography, if users are willing pay for the final products. It will be a wise decision if the state governments decide to produce such databases, engage private entrepreneurs and provide them adequate security cover to carry out the work. Perhaps the Government of Tamil Nadu can take the lead, as it has done in so many cases.